LABORATORY INVESTIGATIONS IN
CELL & MOLECULAR BIOLOGY

LABORATORY INVESTIGATIONS IN CELL & MOLECULAR BIOLOGY

THIRD EDITION

Allyn Bregman

Professor of Biology
State University of New York—College at New Paltz

WILEY

JOHN WILEY & SONS
New York • Chichester • Brisbane • Toronto • Singapore

COVER PHOTO: Intermediate stages of spermiogenesis in the grasshopper *Melanoplus femur-rubrum*. The oval and elongate structures are the heads of differentiating sperm, while the bundle of strands at the far right contains the heads of mature sperm. Spermiogenesis is discussed in Project 16.

Library of Congress Cataloging in Publication Data:
Bregman, Allyn
  Laboratory investigations in cell & molecular biology / Allyn
Bregman. —3rd ed.
    p.   cm.
  Rev. ed. of: Laboratory investigations in cell biology. 2nd ed.
©1987.
  Includes bibliographical references.
  ISBN 0-471-51155-2
  1. Cytology—Laboratory manuals. 2. Molecular biology—Laboratory
manuals.   I. Bregman, Allyn A. Laboratory investigations in cell
biology.   II. Title.
  [DNLM: 1. Cytological Technics. 2. Cytology—laboratory.
3. Molecular Biology—laboratory manuals.   QH 583.2 B833L]
QH583.2.B74   1990
574.87'078—dc20
DNLM/DLC
for Library of Congress   89-22751 CIP

CONTENTS

PREFACE

Since publication of the first edition in 1983, this lab manual has undergone many changes that have broadened the topical coverage. To reflect the increased number of projects in the area of molecular biology, the third edition has been given a new title. The manual is still designed for mid-level cell biology courses, and the topical sequence continues to parallel the sequence found in most cell biology texts. The early projects deal with biochemistry and cytochemistry, the middle projects focus on organelles and their physiology, and the later projects explore more advanced molecular topics.

Two new investigations deal with restriction mapping. In Project 17, the recognition site for a restriction enzyme is mapped within the genome of phage lambda. Project 18 uses a restriction mapping strategy to probe the organization of highly repetitive DNA in the bovine genome. An additional option in the area of protein analysis is provided by Project 7, which is a comparative study of the serum protein fractions in four mammals (calf, goat, guinea pig, horse). Also new in the third edition is a cytological investigation of constitutive heterochromatin; Project 19 employs a simple and reliable method to C-band fixed slides of human chromosomes. Two other investigations (Projects 4 and 16) are based on the former projects on mitosis and meiosis, but with less emphasis on the division stages per se. In Project 4, the emphasis is on chromosome morphology as observed in Feulgen-stained root tips. Project 16 has a new section on spermiogenesis (in the grasshopper), which provides an excellent example of a differentiation process.

The projects have been developed so as to make them as workable as possible. They can be carried out with a minimum of elaborate equipment and with a reasonable amount of preparation time. The materials are readily available and, wherever possible, botanical sources and lyophilized extracts have been chosen. The choice of materials was also guided by safety considerations. For example, DNA bands in electropherograms are stained with a conventional dye instead of with ethidium bromide, and the electrophoresis of proteins is carried out with cellulose acetate and agarose gel plates instead of with polyacrylamide gels. A complete Instructor's Guide follows the last project.

As in previous editions, each project opens with a clearly written introduction, stating the objectives and explaining the theory that relates directly to the lab experience. Detailed procedural steps permit students to make acceptable preparations without the need for constant directions from the instructor. The exercises and questions draw upon the observations, so that students are compelled to examine their data and relate them to the biological principles being investigated. Data sheets and graph paper are included to facilitate collection and analysis of the data. All bench work can be completed in a 3-hour lab session; some projects require a second session for observations.

The following contributions to the third edition deserve acknowledgment. My wife, Sybil, has provided invaluable assistance, critically reading the manuscript at every stage. I am grateful to Uzi Nur (Department of Biology, University of Rochester) for his suggestions on the section on spermiogenesis. I would like to thank Linda Larrabee for the editorial support she gave to the third edition. Grateful acknowledgment is also due Dennis Sawicki, Executive Editor at John Wiley, and his assistant, JoAnn Spear, who have seen the project through to completion. Both Savoula Amanatidis, Production Supervisor, and Lynn Rogan, Design Department, did a splendid job. Also to be acknowledged are the helpful reviews of the new and revised projects by:

Catherine Machalinski (Division of Biological Sciences, University of Georgia, Athens, GA)

David J. Houck (Biological Sciences Department, SUNY, Cortland)

Charles H. Woodward (Division of Science and Mathematics, Shepherd College, Shepherdstown, WV).

The comments on the prospectus by Lester Turoczi (Department of Biology, Wilkes College, Wilkes-Barre, PA) are also appreciated.

Allyn Bregman
New Paltz, NY

MICROSCOPY

For most biologists, the microscope is an important tool, but for the cell biologist, it is indispensable! From the earliest observations of protozoans by Antony van Leeuwenhoek (1632–1723) to today's sophisticated research with fluorescence and interference optics, the microscope has remained the basic instrument for probing the structural basis of life. The microscopes of van Leeuwenhoek are considered *simple microscopes*, for there is but a single lens, as in a magnifying glass. Much greater magnification is achieved with a two-lens system, a *compound microscope*. The objective of this project is to learn the proper use of the compound microscope as you examine some plant cells, a protist, and bacteria.

THE COMPOUND MICROSCOPE

The construction of a modern bright-field microscope is shown in Figure 1.1. The *illuminator* in most modern microscopes is built into the base. It consists of a high-intensity bulb, transformer, lens system, and adjustable *field diaphragm*, or *field iris*, that controls the size of the illuminated field of view. Light passes from the illuminator window to the *substage condenser*, an adjustable lens system that focuses the light on the object. In the lower part of the substage condenser there is an *iris diaphragm*, which controls the diameter of the light beam entering the condenser. The two lens systems that are involved in image formation are the *objective*, which is close to the object, and the *ocular*, or *eyepiece*, which is close to the eye. The objective produces an enlarged and inverted projection of the object on the other side of the lens. This first image (a real image) serves as the object for the ocular. The ocular produces a final image (a virtual image) that is greatly enlarged and still inverted.

Magnification. The magnifications in a typical set of objectives are $10\times$ (low power), $40\times$ (high-dry), and $100\times$ (oil-immersion). Most eyepieces have a magnification between $8\times$ and $12.5\times$. The *total magnification* is the product of the magnifications of the objective and ocular. Thus, the magnification with a $100\times$ objective and $10\times$ ocular is $1000\times$.

Resolution and Numerical Aperture. In addition to magnifying the image, the microscope resolves detail, i.e., reveals the subunits of which an object is composed. The resolving power of a microscope is specified by the *limit of resolution* (*l.r.*), defined as the smallest distance by which two neighboring points can be separated and still be discerned as separate entities. The l.r. of the naked eye is about 0.1 mm, whereas that of a high-quality light microscope is about 0.2 μm, a 500-fold improvement. Thus, if two dotlike structures in a microscope specimen are separated by less than 0.2 μm, they will appear as a single, oblong dot.

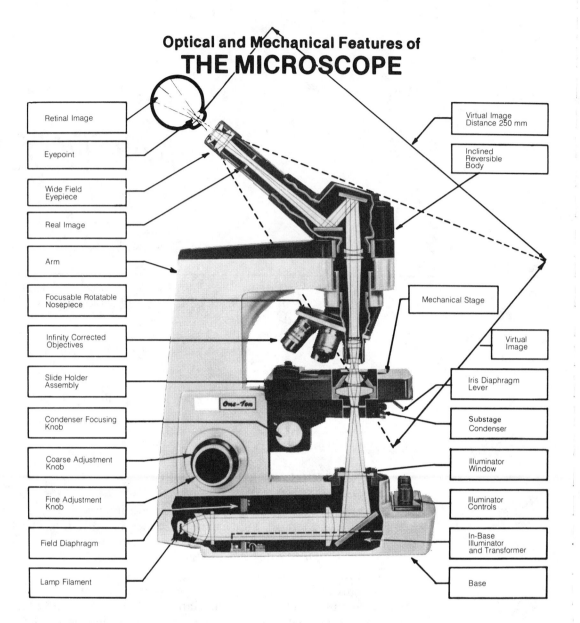

Optical and Mechanical Features of THE MICROSCOPE

Retinal Image

Eyepoint

Wide Field Eyepiece

Real Image

Arm

Focusable Rotatable Nosepiece

Infinity Corrected Objectives

Slide Holder Assembly

Condenser Focusing Knob

Coarse Adjustment Knob

Fine Adjustment Knob

Field Diaphragm

Lamp Filament

Virtual Image Distance 250 mm

Inclined Reversible Body

Mechanical Stage

Virtual Image

Iris Diaphragm Lever

Substage Condenser

Illuminator Window

Illuminator Controls

In-Base Illuminator and Transformer

Base

FIGURE 1.1
Components and optical path in a modern microscope. (Courtesy of Cambridge Instruments.)

The limit of resolution for a microscope is calculated from the *Abbe equation*,

$$\text{l.r.} = \frac{0.61\lambda}{\text{N.A.}}$$

where λ is the wavelength of the light used to view the object and *N.A.* is the *numerical aperture*, a measure of the cone angle of light entering the objective lens (Fig. 1.2). As can be seen from the Abbe equation, the limit of resolution decreases as the N.A. increases. Thus, the larger the N.A., the greater is the resolving power. The increased resolving power results from the larger numerical aperture reducing the *diffraction pattern*, which is the fringe of dark and light rings around object borders. When the fringes become too great, it becomes impossible to discern adjacent objects as separate entities. The highest-quality objectives have an N.A. of 1.40. The lowest possible limit of resolution attainable with a light microscope accrues when a 1.40-N.A. objective is used with a violet filter ($\lambda = 400$ nm). According to the Abbe equation, the limit of resolution equals 0.61(400 nm)/1.40, or 0.17 μm.

In practice, however, the microscope may be operating at a lower N.A., resulting in a somewhat larger l.r., about 0.25–0.40 μm. The *working N.A.* of the microscope equals *n sine* θ. The angle θ is one-half the cone angle of light entering the front lens of

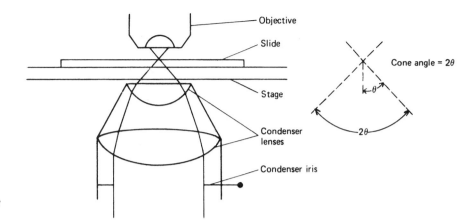

FIGURE 1.2
Light path from the substage condenser to the front lens of the objective.

the objective (Fig. 1.2). The cone angle is controlled by the iris diaphragm of the substage condenser. Opening the *condenser iris* increases the cone angle of light. The condenser iris must be sufficiently open for the working N.A. to equal the N.A. engraved on the barrel of the objective, which is the highest value attainable for that lens. Usually, the condenser iris is partially closed to improve image contrast. When this adjustment is made, the cone angle and, hence, the working N.A. are reduced. Thus, an objective that has an N.A. of 1.25 engraved on the barrel is usually operating at an N.A. that is close to 1.0. The working N.A. is also a function of n, the refractive index of the medium between the coverslip and the objective lens. With the low power and high-dry objectives, the medium is air ($n = 1.0$), but with the oil-immersion objective, a drop of *immersion oil* ($n = 1.515$) is placed between the coverslip and the lens. The refractive index of immersion oil is approximately the same as that for glass so that light is not refracted as it passes from the object to the objective.

Lens Characteristics. All lenses have aberrations. Two types of aberrations that are of particular concern in microscopy are *chromatic* and *spherical*. Left uncorrected, chromatic aberration results in color fringes around objects. When spherical aberration is present, the image may be fuzzy, and straight lines may appear curved. Your microscope is probably equipped with *achromat objectives*, which are corrected chromatically for two colors (blue and red) and spherically for one (green). Another important lens correction is for *curvature of field*. With an uncorrected objective, the entire field is not in focus at the same focal setting. Objectives designed to produce a *flat field* are designated with the prefix *plan*.

Several significant optical characteristics of a microscope are determined by the N.A. and the magnification of its lenses. These are summarized in Table 1.1 for a typical set of planachromat objectives. The *maximum useful magnification* is the limit that increased magnification can improve our ability to resolve detail. Any greater magnification will not improve our ability to discern but only reduce the quality of the

TABLE 1.1
Relationship of N.A. and objective magnification to maximum useful magnification, working distance, depth of field, and diameter of field. Data are for Reichert planachromat objectives used with 10× wide field oculars in a Series 110 microscope. (Data courtesy of Cambridge Instruments.)

Objective lens	N.A.	Maximum useful magnification (1000 × N.A.)	Working distance	Depth of field	Diameter of field
Plan Achro 10×	0.25	250×	9.1 mm	8.8 μm	2.0 mm
Plan Achro 20×	0.50	500×	1.4 mm	2.2 μm	1.0 mm
Plan Achro 40×	0.66	660×	0.5 mm	1.3 μm	0.5 mm
Plan Achro 100× (oil)	1.25	1250×	0.1 mm	0.5 μm	0.2 mm

final image. Depending on lens quality, the maximum useful magnification can be as great as 1000 times the N.A. Another characteristic of the objective is its *working distance*, defined as the clearance between the top of the coverslip and the lowest part of the objective when the specimen is in focus. As the N.A. increases, the working distance decreases dramatically in order to accommodate the larger cone angle of light. The *depth of field*, which is the specimen thickness that is in focus at one focal setting, also decreases sharply with increasing N.A. Finally, there is the *diameter of the field*, which is inversely proportional to the total magnification; with a doubling of magnification, the diameter of the field is halved.

Köhler Illumination.   To achieve the highest intensity and most uniform illumination from a nonuniform source, such as the filaments of a bulb, the microscope components must be adjusted for *Köhler illumination*. The two basic steps in the method, which was introduced by August Köhler in 1893, are (1) focusing the bulb filaments at the plane of the condenser iris and (2) focusing an image of the field iris at the plane of the object. In microscopes with an in-base illuminator, the illuminator lens system is positioned at the factory so that step 1 need not be performed. You will carry out step 2 by adjusting the height of the substage condenser. After obtaining a bright and uniformly illuminated field, you will adjust the aperture of the condenser iris to obtain the best compromise of resolving power and contrast.

Contrast, Brightness, and Filters.   Contrast is controlled by the condenser iris. Generally, the condenser iris should be open so that about two-thirds the diameter of the back focal plane of the objective is filled with light. With a lightly stained or unstained specimen, the condenser iris may have to be further closed; resolving power will be reduced, but contrast and depth of field will be dramatically improved. Image contrast is also enhanced by using an appropriate colored filter. In the calculation of the lowest possible l.r., it was assumed that a violet filter was used, but, in practice, a *green* filter (λ = 500–600 nm) is placed in the light path. The green illumination is nearly complementary to the red and purple hues produced by many biological stains and, thereby, increases contrast. In addition, the human eye is most sensitive to green light, and objectives are corrected for aberrations in this spectral region.

Brightness is controlled by the transformer setting and by *neutral density filters*, which reduce light intensity without altering the color quality of the illumination. The brightness should never be reduced by closing the condenser iris or by lowering the substage condenser since there will be a concomitant loss in resolution. Bulb life can be prolonged by using the lowest transformer setting and by not switching the illuminator on and off repeatedly. The illuminator should be left on until all observations have been made for that laboratory session.

Microscope Technique.   All modern microscopes are *parfocal* and *parcentral*. The term parfocal indicates that the focus does not change appreciably when the objective is changed; only slight touching up of the focus with the fine adjustment knob is required. The term parcentral indicates that the center of the field does not change when the objective is changed. Before the objective is changed, the object should be centered in the field.

Always start focusing with low power first, checking that the coverslip is on the upper surface of the slide so that the higher-power objectives will not hit the slide. To obtain a good image with the high-dry objective, a coverslip must be used, and it should be of No. 1½ thickness (0.17–0.18 mm). Oil-immersion objectives are so designated and *must* be used with immersion oil between the front lens and the coverslip. The oil-immersion objective is not as sensitive to coverslip thickness and can even be used without a coverslip. Keep in mind that the oil-immersion objective has a very short working distance and can easily impact on the slide. When the objective is switched back to high-dry, it is necessary to wipe the immersion oil from the coverslip. After all observations have been made, the oil-immersion lens should be wiped with clean lens paper. *Wide field eyepieces* are a great advantage since they produce a large field of view and place the eyepoint (point where the exiting light rays converge) far enough from the lens to provide adequate clearance for eyeglasses. Actually, since the microscope's optics can correct for nearsightedness and farsightedness, eyeglasses need not be worn unless there is severe astigmatism. So that your eyes are at the level of the eyepieces and viewing is comfortable, it is important that the chair height be properly adjusted.

Image quality is dramatically impaired by dust and body oils on the lenses. The ocular can be cleaned with lens paper moistened with distilled water. When the microscope is not in use, it should be covered to keep dust off the lenses. Finally, the microscope should always be carried in an upright position, with two hands.

PHASE-CONTRAST MICROSCOPY

When detailed study of unstained specimens is required, we generally rely on the phase-contrast microscope. With phase optics, details of a low-contrast object are made apparent because of a phenomenon known as *interference*. If two light waves arrive at some point with their crests and troughs in phase, there is *constructive interference* and a great increase in brightness. Conversely, if the two waves are completely out of phase, there is *destructive interference*; the two waves cancel each other out, and the intensity is zero. When light waves pass through a typical unstained cell preparation, they are slowed, so that as they emerge from the specimen they are retarded by about one-fourth of a wavelength ($\frac{1}{4}\lambda$). The special light path in a phase-contrast microscope (Fig. 1.3) selectively retards these light waves yet another $\frac{1}{4}\lambda$. Compared to the light waves that did not strike the object, they are now out of phase by $\frac{1}{2}\lambda$ (inset at upper left of Fig. 1.3). The visual effect of the $\frac{1}{2}\lambda$ interference is a dark object contrasted against a lighter background.

The contrast enhancement in a phase-contrast microscope is effected by a special substage condenser and phase objectives. The substage condenser contains a rotatable turret mount with several *annular diaphragms* (bottom of Fig. 1.3), one corresponding to each phase objective. One position has no annulus so that bright-field illumination can be obtained. Each annulus directs a hollow cone of light to the specimen. The light waves that do not strike the object are termed the *direct rays*, for they are not diffracted and continue on path to the objective. Those light waves that do strike the object are retarded by $\frac{1}{4}\lambda$ and are diffracted from their path. At the back of the phase objective there is a *phase plate*, containing a ring-shaped disk that accommodates the direct rays and allows them to pass through without retardation. The ring also contains absorbing material to reduce the intensity of the direct rays. The remainder of the phase plate is composed of material that retards light waves by $\frac{1}{4}\lambda$. The *diffracted rays*, which strike all regions of the phase plate, are thereby retarded by a total of $\frac{1}{2}\lambda$ ($\frac{1}{4}\lambda$ by the specimen plus $\frac{1}{4}\lambda$ by the phase plate). To achieve maximum contrast, each annulus must be aligned with its respective phase plate. A *phase telescope* or built-in *telescopic lens* is used for viewing both the annulus and phase plate during the alignment procedure.

PROCEDURES

FAMILIARIZATION WITH MICROSCOPE COMPONENTS

1. Familiarize yourself with your microscope by identifying the components labeled in Figure 1.1. Trace the light path from the illuminator to the eyepiece.
2. Note the magnification of each objective and of the ocular on your microscope. Then compute the *total magnification* with each objective and enter the values in the table on page 11.

BRIEF DESCRIPTION OF PHASE

Essentially, phase contrast is produced by the combination of (a) an annular diaphragm located below the condenser that directs a hollow cone of light through the specimen and (b) a phase plate at the back focal plane of the objective. Some of the light passing through the transparent specimen is diffracted by slight differences in optical path (refractive index × thickness) and moves so as to be distributed over the whole aperture of the objective. The balance of the light passes directly through the specimen as a cone of concentrated light toward coincidence with the "ring" of the phase plate. The phase plate alters the intensity and phase relationships of the diffracted and direct light so that when they recombine to form an image, invisible specimen optical path differences are converted into visible light-intensity differences.

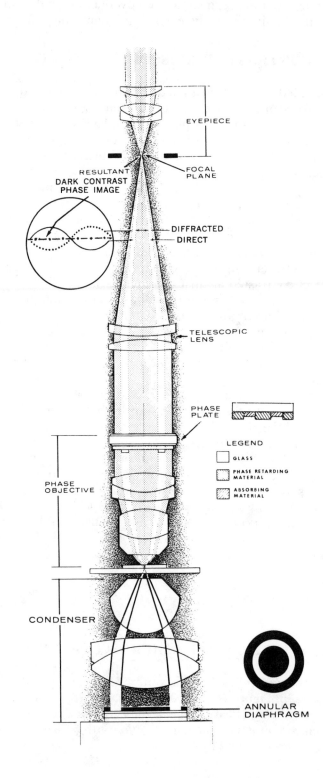

FIGURE 1.3
Optical path in a phase-contrast microscope. *(Courtesy of Cambridge Instruments.)*

3. Note the N.A. of each objective and compute the best limit of resolution attainable with each. Assume that a green filter (λ = 550 nm) is used and that the working N.A. with the oil-immersion objective is 1.0. Enter these values in the table on page 11.
4. Determine the maximum useful magnification and the highest-power ocular to be used for your microscope. Enter both values below the table on page 11.

OBTAINING KÖHLER ILLUMINATION WITH AN IN-BASE ILLUMINATOR

1. Turn on the illuminator, using the lowest transformer setting.
2. By turning the condenser focusing knob, raise the substage condenser to the top of its travel. Open the field iris and the condenser iris all the way.
3. Place a slide with a stained specimen on the stage, checking that the coverslip is on the upper surface. Position the slide so that the specimen is in the light path.
4. Using the 10× objective, bring the object into focus. Start with the objective close to the slide and, with the coarse adjustment knob, increase the distance between the objective and the slide until the object is in focus. Touch up the focus with the fine adjustment knob. If the field is uncomfortably bright, place a neutral density filter over the illuminator window.
5. In a binocular microscope, an accommodation must be made for differences in interpupilary distance; move the ocular tubes closer together or farther apart until both eyes see the image simultaneously. To adjust for differences in the vision of the two eyes, bring the object into sharp focus with the fine adjustment knob while looking through the eyepiece in the fixed-length ocular tube, which is usually the right one. The left eye should be closed or covered with a piece of paper. Now, with the right eye closed or covered, focus the image for the left eye by rotating and, thereby, changing the length of the left ocular tube.
6. Close the field iris almost all the way. Then slightly lower the height of the substage condenser until there is a sharp image of the inner leaves of the field iris superimposed on the image of the object.
7. Center the image of the field iris with the centering screws on the illuminator housing or, if it is not adjustable, with the centering screws on the substage condenser mount. After it is centered, open the field iris until the light just completely fills the field of view. The result of steps 6 and 7 is a uniformly illuminated field.
8. Remove an eyepiece and look down the ocular tube. Close the condenser iris until about two-thirds the diameter of the back focal plane of the objective is filled with light.
9. Replace the eyepiece. The microscope is now properly adjusted for the 10× objective.
10. Change to the high-dry objective and touch up the focus with the fine adjustment knob. Adjust both iris diaphragms as follows. Close the field iris until the light just completely fills the field of view and, if necessary, touch up the focus and centering of the field iris. Open the condenser iris until about two-thirds the diameter of the back focal plane of the objective, observed through an empty ocular tube, is filled with light. Replace the eypiece. The microscope is now properly adjusted for the high-dry objective.
11. Partially rotate the nosepiece, place a drop of immersion oil on the coverslip, and then bring the oil-immersion objective into place. Touch up the focus with the fine adjustment knob. Close the field iris until the light just completely fills the field of view. Generally, the field is so small with the oil-immersion objective that the field iris will be closed to its smallest aperture. The field iris need not be focused if this was previously done for the high-dry objective. Open the condenser iris until about two-thirds the diameter of the back focal plane of the

objective, observed through an empty ocular tube, is filled with light. Replace the eyepiece. The microscope is now properly adjusted for the oil-immersion objective. For routine visual observations, the eyepiece need not be removed to adjust the condenser iris every time the objective is changed; one simply adjusts it until there is good contrast.

CALIBRATION OF AN OCULAR MICROMETER

The ocular of a microscope can accommodate a glass disk with markings (pointer, cross hairs, etc.) that are in focus with the object. For object measurement, use an *ocular micrometer*, a disk with calibrated divisions.

1. With the ocular micrometer in place, focus on the calibrations of a stage micrometer slide using the low power objective.
2. Rotate the eyepiece until the ocular micrometer scale is aligned with the stage scale.
3. Determine the length of each division (in micrometers) by dividing the total length subtended on the stage scale by the total number of divisions on the eyepiece scale. Most ocular micrometers have a total of 50 or 100 divisions. Enter the values in the table on page 11.
4. Carefully change to the high-dry objective and determine the length (in micrometers) of each division. Enter the values in the table on page 11.
5. Do *not* attempt to focus on the calibrations with the oil-immersion objective, which could easily impact on the thick micrometer slide. Instead, calculate the length of each division from the value obtained at low power. Divide the value for micrometers per division by 10 (assuming a $100\times$ oil-immersion objective and a $10\times$ low power objective, i.e., $100/10 = 10$). Enter the value in the table on page 11.

WET MOUNTS

In a wet mount, a piece of tissue is placed on a clean slide with a drop of water, stain, or reagent. A coverslip is then gently lowered on the preparation.

Onion Epidermis.
1. Add a small drop of water to the center of a clean slide.
2. Cut a fresh onion in half and remove a layer.
3. Using pointed forceps, strip a small piece of epidermis from the concave surface of a layer and place it on the drop of water, being careful that it does not fold over on itself. Add a drop of water and a coverslip.
4. Examine the cells with the low power and high-dry objectives. Answer the first four questions on page 11.
5. Add one or two drops of *Lugol's iodine solution* at one edge of the coverslip and draw it through by touching a piece of paper towel to the opposite edge of the coverslip. Examine the stained cells (at the edge where the stain was added) with the low power and high-dry objectives, varying the opening of the condenser iris until the nucleus is clearly visible. How does the adjustment of the condenser iris differ from that used for the unstained cells? Answer on page 11.
6. Measure the diameter of 10 nuclei, selecting cells in which the nucleus appears round. Calculate the *mean* ($\overline{X}$), defined as $\overline{X} = \Sigma X/N$, where ΣX is the sum of the individual values and N is the total number of measurements. Enter the 10 values and the mean on page 12.
7. On the blank paper provided at the end of this project (p. 13), draw a typical onion epidermal cell, labeling the visible cell structures. Every drawing done

should have a legend containing the name of the organism (here, *Allium cepa*), the cell type, the stain or technique used in its preparation, and the magnification of the drawing, or *plate magnification*, which is (size of object in the drawing)/(actual size of object). For example, if the length of a cell in a drawing is 70 mm and its actual length is 35 μm, the plate magnification is (70 mm/35 μm) = 2000×. Don't make the mistake of recording the magnification at which the object was viewed!

Elodea Leaf.

1. Using a healthy sprig of *Elodea*, prepare a wet mount of a leaf in a drop of water. The upper surface of the leaf should be facing up.
2. Before focusing on the preparation and with the *low power* objective in place, turn the coarse and fine adjustment knobs in one direction and then in the other direction as you watch the movement of the nosepiece or stage from the side. Determine whether the focal plane moves up or down when you turn the focus knobs away from you and toward you.
3. With the green filter removed, examine the cells under low power and high-dry. Determine how many cell layers there are and which layer is composed of larger cells. Enter your answers on page 12.
4. Using the high-dry objective, determine the thickness of the *Elodea* leaf as follows. Record the line division on the fine adjustment knob when the upper leaf surface just comes into focus and again when the lower leaf surface just starts to go out of focus. The number of divisions multiplied by the distance/division (usually 1 μm/division) gives the vertical distance traversed. Repeat twice, recording the three values and the mean on page 12.
5. Scan the preparation for cells exhibiting *cyclosis*, or *cytoplasmic streaming*, and determine the approximate rate in one cell. To obtain the rate of cyclosis, measure the length of the cell and the time required for a chloroplast to move from one end of the cell to the other. Enter the data on page 12.
6. Draw a typical *Elodea* leaf cell, labeling the cell wall, cytoplasm, vacuole, chloroplasts, and nucleus (if visible). Record the plate magnification.

Euglena.

1. On a clean slide, place a drop of the *Euglena* culture and add a coverslip. Observe several specimens as they swim across the field. Look closely for all the ways in which they are moving and then answer the question on page 12. Locate a nonswimming specimen and observe its characteristic movement, termed *euglenoid movement*. Describe euglenoid movement on page 12.
2. Prepare a fresh wet mount using one drop of the *Euglena* culture and a small drop of *Protoslo* (Carolina Biological Supply Co.), a viscous liquid used to slow swimming protozoa. Mix with a toothpick, add a coverslip, and examine under low power and high-dry. Identify the anterior (leading) and posterior (trailing) ends of the *Euglena* and determine from which end the flagellum emerges. Remember to close the condenser iris to enhance contrast. Enter your answer on page 12.
3. Examine the internal structure of the slowed *Euglena*. Draw a *Euglena*, labeling all visible structures and recording the plate magnification.

PREPARED SLIDES

Zea Stem, Cross Section.

1. Scan the stained preparation under low power and identify the numerous *vascular bundles* scattered near the periphery of the stem.
2. Center a vascular bundle in the field, add a drop of immersion oil, and switch to

the $100\times$ objective. Identify the large *xylem vessel elements*, the cell walls of which are stained red.

3. With the green filter in place, measure the thickness of the cell wall in three vessel elements (in regions that are not in contact with the wall of another vessel element). Enter the values and the mean on page 12.
4. Examine the preparation without the green filter and compare the image to that observed *with* a green filter. Explain on page 12.

Bacteria.

1. Scan a stained preparation under high-dry to locate an area with a concentration of bacteria.
2. Switch to oil-immersion and examine the morphology of the cells.
3. Draw a representative cell or group of cells for each bacterial species supplied. Next to each drawing, give the dimensions and plate magnification.

OPTIONAL PROCEDURE

PHASE-CONTRAST MICROSCOPY

1. Turn on the illuminator. An intense light source is required for phase microscopy, and colored filters are not used for visual observations.
2. Rotate the turret on the phase condenser to the "open" position, i.e., to the position without any annular diaphragm.
3. With a stained specimen on the stage and the $10\times$ phase objective in place, adjust the microscope for Köhler illumination (through step 7, p. 7).
4. Open the iris diaphragm of the substage condenser; the condenser iris is left fully open for *all* phase microscopy.
5. Rotate the turret to bring the $10\times$ annulus into position.
6. Replace the right eyepiece with a phase telescope and adjust its length until the annulus and phase plate are in sharp focus. In some microscopes, a telescopic lens is built into the body and must be swung into position.
7. While looking through the phase telescope, turn the centering wrenches in the turret and align the $10\times$ annulus so that it is concentric with the ring in the phase plate of the objective.
8. Center the other annuli for their respective objectives. When changing phase objectives, check that the field iris is centered and that its image is focused at the same plane as the object; if the condenser height is incorrectly adjusted, the annulus image will be partially colored. Replace the phase telescope with the regular eyepiece.
9. Place a wet mount of unstained onion epidermis and/or *Euglena* on the stage and examine with the low power and high-dry phase objectives. What structures are more clearly delineated with phase than with bright-field optics? Enter your answers on page 12.

REFERENCES

Curry, A., Grayson, R. F., and Kosey, G. R. 1982. *Under the Microscope.* Van Nostrand Reinhold, New York.

Locquin, M. and Langeron, M. 1983. *Handbook of Microscopy.* Butterworths, London.

Spencer, M. 1982. *Fundamentals of Light Microscopy.* Cambridge Univ. Press, Cambridge, England.

OBSERVATIONS AND QUESTIONS

FAMILIARIZATION WITH MICROSCOPE COMPONENTS

	Magnification	Total magnification	N.A.	l.r.
Low Power				
High-dry				
Oil-immersion				
Ocular		—	—	—

Maximum useful magnification _____

Highest-power ocular to be used _____

CALIBRATION OF AN OCULAR MICROMETER

	Length subtended on stage scale (mm)	Length, μm/division
Low power		
High-dry		
Oil-immersion	—	

WET MOUNTS

Onion Epidermis.
How many cell layers are there? _____

What cell parts are visible? _____
How must the condenser iris be adjusted to observe the nucleus?

_____

Why? _____
How has the adjustment of the condenser iris changed when observing the stained preparation?

_____

_____ _____ _____ _____ _____ _____ _____ _____ _____ _____ _____

$\overline{X}$ = _____ μm

Elodea Leaf.
Number of cell layers _____

Layer with larger cells _____

Leaf thickness (μm) _____   _____   _____ $\overline{X}$ = _____ μm

Length of cell = _____ μm; time required for movement = _____ sec

Rate of cyclosis = _____ μm/sec

Euglena.
What other motion do swimming specimens exhibit besides forward motion?

_____

Describe the motion of nonswimming *Euglena*, i.e., euglenoid movement.

_____

Does the flagellum emerge from the anterior or posterior end of the organism?

_____

PREPARED SLIDES

Zea Stem, Cross Section.
Cell wall thickness (μm) _____   _____   _____ $\overline{X}$ = _____ μm
How has the image improved with a green filter?

_____

PHASE-CONTRAST OBSERVATIONS

What structures are more apparent in onion epidermis?

_____

What structures are more apparent in *Euglena*?

_____

CELLULAR CARBOHYDRATES

INTRODUCTION

Carbohydrates and macromolecules that contain sugar moieties perform a variety of functions in the cell and in the organism. The glycogen in animals and the starch in plants serve as energy stores. Other carbohydrates, such as chitin and cellulose, have structural roles in animals and plants, respectively. Oligosaccharides are components of the glycoproteins and glycolipids found in the plasma membrane. These sugar chains, which always face the cell's exterior, stabilize the position of glycoproteins and glycolipids within the membrane, function in cell adhesion, and confer immunological specificity to the cell surface.

Every class of macromolecule can be localized in cells with specific cytochemical reactions. The term *cytochemistry* can refer to any methodology that probes the chemical nature of the cell but usually is reserved for specific staining reactions and subsequent microscopic analysis. An important cytochemical method used to identify cellular carbohydrates is the *periodic acid–Schiff (PAS) reaction*. The reaction stains insoluble polysaccharides (glycogen, starch, cellulose), mucopolysaccharides, mucoproteins, glycoproteins, and glycolipids. The objectives of this project are to carry out the PAS reaction on fixed blood smears and to localize PAS-positive material in white blood cells and in other formed elements of the blood.

BLOOD CELL TYPES

Histologists frequently use *Wright–Giemsa stain* for the identification and study of human blood cells. Wright–Giemsa stain is a mixture of methylene blue, methylene azure, and the eosinates of both. The azures act as bases and stain the *basophilic* (base-loving) elements of the cells blue, while the eosins behave as acids and stain the *acidophilic* (acid-loving) structures red. Since there is a combination of dyes and a varying affinity for each, the various parts of the cell are stained in hues of pink, purple, blue, and red. After Wright–Giemsa staining at pH 6–7, the structures in blood cells have characteristic colors, and each cell type is readily identifiable.

Figure 2.1 is a diagram of the various human blood cell types as they appear in a fixed smear. There are two classes of cells, red blood cells, or *erythrocytes*, and white blood cells, or *leukocytes*. The leukocytes are placed into two groups, the *granulocytes*, which have conspicuous cytoplasmic granules, and the *agranulocytes*, which lack them. The granulocytes include the *neutrophils*, *eosinophils*, and *basophils*. The agranulocytes include the *lymphocytes* and *monocytes*. In addition to these cell types there are the *platelets*, which are small cell fragments.

Erthrocytes. The erythrocyte is the most frequent cell type. It is also the smallest blood cell, about 7.5 μm in diameter and 1.9 μm in thickness at the periphery. Human erythrocytes lack a nucleus and are biconcave. In Wright–Giemsa–stained preparations, they appear pink with a central region staining lighter because of the concavity. They lack any internal organelles but are filled with hemoglobin for the function of oxygen transport.

Neutrophils.   Neutrophils are the most frequent type of white blood cell, comprising 55–75% of the leukocyte population. In stained fixed smears, neutrophils are round, 9–12 μm in diameter, and easily recognized by their dark purple, multilobed nucleus. Most are *segmented neutrophils*, in which the nucleus has 3–5 lobes connected by a thin strand of chromatin. The young neutrophils, which do not have their nuclei divided into lobes, are called *band neutrophils*. Because of the varied nuclear morphology, neutrophils are also termed *polymorphonuclear leukocytes*. The cytoplasm is pale pink and contains small, very lightly stained *specific granules*. There are also a few larger, lavender *azurophil granules*. Some of the larger granules are *lysosomes*, which function in intracellular digestion. They are membrane-bound vesicles that contain *acid hydrolases*, i.e., hydrolytic enzymes with acid pH optima. Neutrophils are the body's first defense against invading microorganisms. At a wound site, for example, they leave the bloodstream and engulf bacteria that may be present.

Eosinophils.   Eosinophils are named for the tendency of their cytoplasmic granules to stain with acidic dyes. They comprise only 1–5% of the leukocyte population. Eosinophils are round cells, 10–14 μm in diameter, and have a dark purple nucleus that usually has two large lobes connected by a thin strand or a band of chromatin. The cytoplasm is distinctive, being filled with many large, reddish orange *eosinophil granules*. The eosinophil granules are true lysosomes since they contain acid hydrolases within a limiting membrane. It appears that eosinophils selectively phagocytize foreign proteins that are complexed with antibodies.

Basophils.   Basophils are named for the tendency of their cytoplasmic granules to stain with basic dyes. They are the least frequent cell type; basophils comprise only 0.5% of the leukocyte population and, hence, may be difficult to find. In fixed smears, they are about 10 μm in diameter. The nucleus, which stains purple, may be round, indented, band, or lobed. The cytoplasm contains numerous large, dark purple *basophil granules*, which serve to distinguish this cell type. The basophil granules are not lysosomes, for instead of acid hydrolases, they contain histamine and heparin. The function of the basophils is not well understood.

Lymphocytes.   Lymphocytes are the second most abundant type of white blood cell, comprising 20–40% of the leukocyte population. They vary considerably in size; most lymphocytes are small, 7–10 μm in diameter, though some may be as large as 15 μm. The nucleus is round and large, often with an indentation. It stains dark purple, and there are usually clumps of chromatin present. The cytoplasm stains blue. In the small lymphocytes there is only a thin layer of cytoplasm surrounding the nucleus. The primary function of lymphocytes is the production of antibodies in the immune response.

Monocytes.   Monocytes are large cells, 12–15 μm in diameter, and comprise 3–8% of the leukocyte population. Many are round or oval, and some have blunt *pseudopods*. The general appearance is that of a large lymphocyte, but the two cell types are distinguishable by their staining. The monocyte nucleus, which is usually round or kidney-shaped, stains lightly and does not have many clumps of chromatin. The monocyte cytoplasm is a dull gray-blue and contains small, lilac-stained granules. Monocytes remain in the bloodstream only 1–2 days; they function mainly in the body tissues where, as *macrophages*, they phagocytize a variety of foreign substances.

Platelets.   In addition to all the foregoing cell types there are the platelets. They are small, irregularly shaped cell fragments that have broken away from larger cells in the

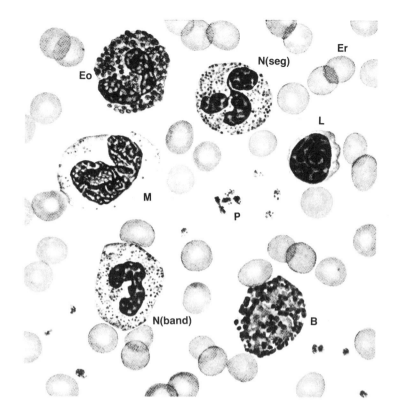

FIGURE 2.1
Human blood cells as they appear in a fixed smear. Er, erythrocytes; N(seg), segmented neutrophil; N(band), band neutrophil; Eo, eosinophil; B, basophil; L, lymphocyte; M, monocyte; P, platelets. (Courtesy of Abbott Laboratories, from L. W. Diggs, D. Sturm, and A. Bell, *The Morphology of Human Blood Cells*, 5th edition, © 1985, p. 5.)

bone marrow. Platelets range in size from 1–4 μm and stain blue or purple. These numerous cell fragments play a major role in the blood clotting process.

PERIODIC ACID–SCHIFF REACTION

The periodic acid–Schiff reaction was first introduced for histological preparations by J. F. A. McManus in 1946. Figure 2.2 shows the steps in the staining of a carbohydrate (glycogen). The first step in the PAS reaction is treatment of the fixed cells with periodic acid (HIO_4). The HIO_4 oxidizes the 1,2-glycol grouping of sugars to a dialdehyde. The preparation is then stained with *Schiff's reagent*, a colorless liquid prepared by treating *basic fuchsin* (a red dye) with sulfurous acid. Schiff's reagent reacts with aldehydes to form colored compounds (purplish red) by a mechanism that is not completely understood. Any cell structures that stain purplish red with the PAS reaction are said to be *PAS-positive*. Although the PAS reaction cytochemically identifies carbohydrates, it also stains some lipids other than glycolipids, specifically, unsaturated lipids and phospholipids. The PAS reaction does not stain ribonucleic acid (RNA) or deoxyribonucleic acid (DNA) as their sugars lack the 1,2-glycol grouping (see Fig. 3.2, p. 25, for the structures of RNA and DNA).

Counterstaining.   Since the PAS reaction does not stain nucleic acids, the nucleus of the cell shows little or no staining. Accordingly, after the PAS reaction, a nuclear *counterstain* is commonly used. You will counterstain with *Harris's hematoxylin*, a mixture of hematein and ammonium aluminum sulfate. The hematoxylin, which has an affinity for chromatin, stains the nucleus dark blue.

Diastase-Treated Control Slide.   In many cytochemical reactions, control slides are used to verify the specificity of the staining. For the PAS reaction, the control slide is treated with the enzyme *diastase*, which digests starch and glycogen. An absence of staining after diastase treatment confirms the presence of glycogen in PAS-positive regions of the blood cells.

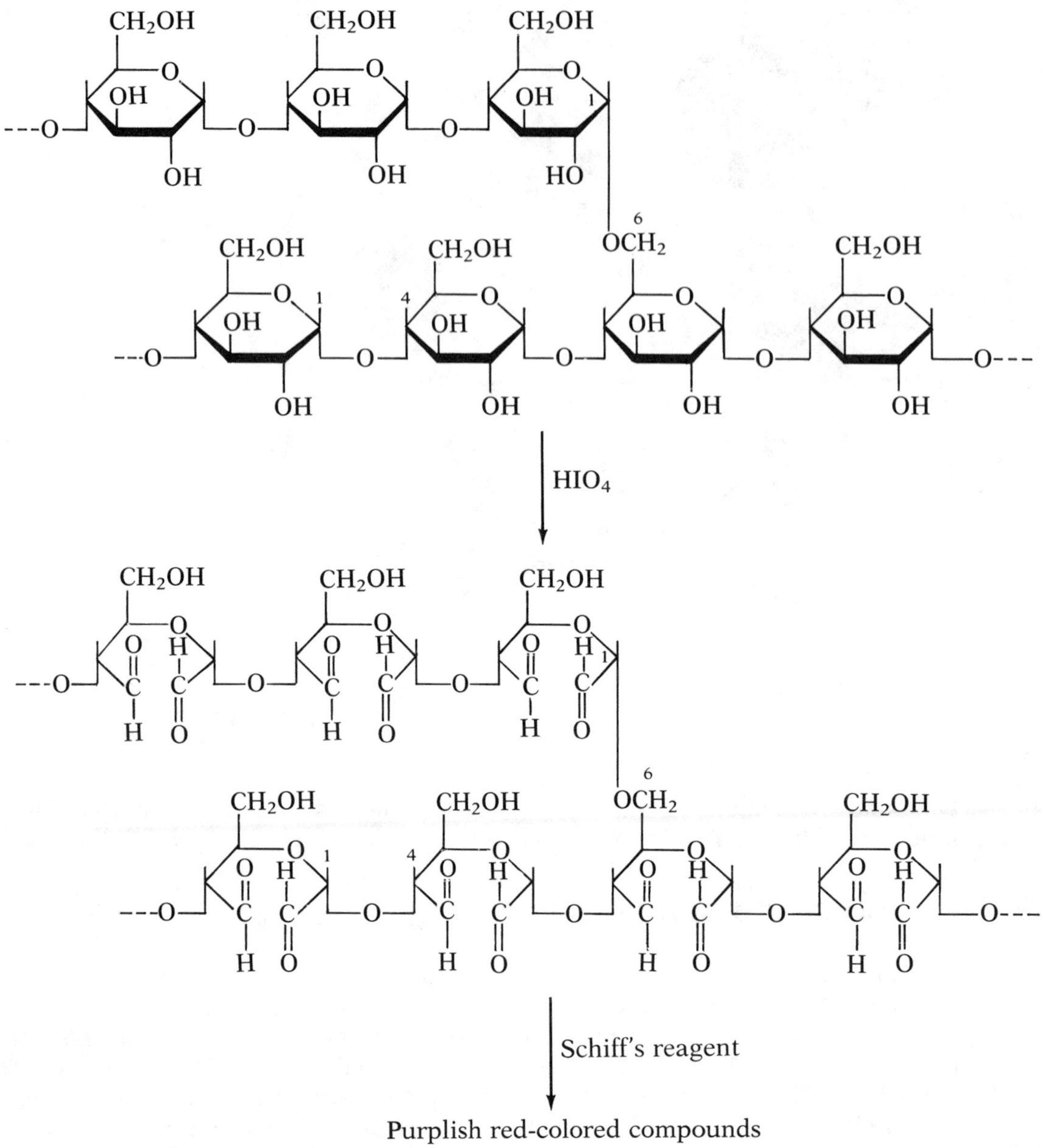

HIO_4

Schiff's reagent

Purplish red-colored compounds

FIGURE 2.2
The steps in the periodic acid–Schiff (PAS) reaction, shown here staining a portion of a glycogen molecule.

Slide Preparation. The procedure selected for the PAS reaction in blood cells is a modification of the method developed by F. G. J. Hayhoe and coworkers in 1964 (see Miale, 1982). You will be provided with two blood smears that have previously been *fixed* in a 9:1 mixture of ethanol:formalin. In cytological preparations, cells must be fixed so that their general morphology and internal structures will be preserved. If the cells were not treated with *fixative*, the hydrolytic enzymes of the lysosomes would be released, and, eventually, much of the cell would be digested. An effective fixative must render cell components insoluble, lest they be washed out during subsequent treatment. It should also prevent subsequent swelling or shrinkage of the cell contents. Often, the fixative improves staining by enhancing the affinity of cell components for the dye molecules. Many fixatives contain an alcohol plus one or more of the following: acetic acid, formalin, chloroform.

For the various steps in the PAS reaction, the fixed blood smears will be transferred to or stored in Coplin jars. As the slides are being processed, the slide surface with the cell preparation must never be allowed to rest against another slide. You will work with two fixed blood smears, one for the PAS reaction and one for the diastase control. After the control slide is treated with diastase, both slides are treated with HIO_4 followed by Schiff's reagent. The staining jar should be kept in the dark since Schiff's reagent deteriorates rapidly in the light. After treatment with Schiff's reagent, the slides are rinsed with a *bisulfite bleaching solution*. This rinse assures that the red coloration is due to the cytochemical reaction and not due to the presence of any basic fuchsin that may have formed during the staining procedure. Subsequently, the slides are counterstained. Only half of each blood smear will be counterstained, so that there will be a portion of each smear stained only by the PAS reaction.

The final step in the procedure is the mounting of a coverslip. The coverslip protects the cell preparation and allows immersion oil to be removed easily. The choice of mounting medium is often dictated by the last solvent used in making the preparation. Sometimes it depends on the stain, as certain stains are leached out of the tissue by some mountants. A popular mountant is *Permount* (Fisher), a synthetic resin that is soluble in xylene. The mountant will dry only after several days at room temperature or overnight in a warm incubator. The slides will be examined before the mountant has dried, so extreme care should be taken not to get the sticky mountant on the oil-immersion lens. When the preparation is completely dry, excess mountant can be removed from the edges of the coverslip with solvent.

PROCEDURES

PERIODIC ACID–SCHIFF REACTION

1. Label the slides (on the frosted end) in pencil with your initials and the staining procedure: one slide, labeled *PAS*, will be subjected to the PAS reaction; the second slide, labeled *D + PAS*, will be treated with diastase prior to the PAS reaction. The pencil markings will also serve to identify the slide surface with the cell preparation.

2. Place the slide labeled *D + PAS* in a 1% solution of diastase in phosphate buffer, pH 6.0, at 24° C, for 30 min. The enzyme treatment and all staining procedures are done in Coplin jars; use a zigzag arrangement of up to nine slides per jar. Always use forceps to place slides in or to remove slides from Coplin jars. During the enzyme treatment, the slide labeled *PAS* can simply be left on the lab bench.

3. Rinse the diastase-treated slides under running tap water for 2 min, in a Coplin jar. Then replace with distilled water and pour it off. Unless otherwise noted, slides do not have to be dried between steps in the PAS reaction.

4. Place both slides (*PAS* and *D + PAS*) in a 1% aqueous solution of periodic acid for 10 min.

5. Rinse the slides under running tap water for 1 min (in a Coplin jar) and then rinse with distilled water, as in step 3.

6. Place the slides in Schiff's reagent* (in a fume hood) for 10 min. The staining jar should be kept in the dark.

7. Remove the slides from the staining jar with forceps and transfer them to a clean Coplin jar. Rinse the slides under running distilled water for several seconds.

8. In a fume hood, replace the distilled water with freshly prepared bisulfite bleaching solution (180 ml distilled water, 10 ml 10% Na or K metabisulfite solution, 10 ml 1 N HCl). Decant after 2 min and repeat this bleaching procedure two more times.

9. Rinse the slides under running tap water for 5 min and then in distilled water. Wipe the back of each slide with a Kimwipe and allow to air-dry in a near-vertical position.

*CAUTION: *Schiff's reagent causes burns; if any spills on your skin or clothing, wash it off immediately with water. Also avoid breathing vapors. Although it is a colorless liquid, it will stain skin, clothing, and other surfaces purplish red. Always keep the staining jar on paper towels or Benchkote.*

10. Place the air-dried slides in the hematoxylin staining solution (counterstain) for 2 min. Each Coplin jar should have only enough of the staining solution (about 15 ml) so that approximately half of the blood smear will be immersed when the slide is placed in the jar, and no more than four slides should be placed in each Coplin jar at one time.

11. After staining, remove each slide from the Coplin jar and immediately rinse under running distilled water for several seconds. Wipe the back of each slide and allow to air-dry in a near-vertical position.

12. When the slides are completely dry, dip each in xylene (in a fume hood) and add a drop of Permount and a coverslip. Position the coverslip so that it will cover all of the smear. Lower the coverslip slowly to minimize air bubbles and then gently press out the excess liquid with a paper towel.

REFERENCES

Diggs, L. W., Sturm, D., and Bell, A. 1985. *The Morphology of Human Blood Cells*, 5th ed. Abbott Laboratories, Abbott Park, IL.

Lillie, R. D. 1977. *H. J. Conn's Biological Stains*, 9th ed., pp. 26–28, 259–266, 489–502. Williams & Wilkins, Baltimore.

Miale, J. B. 1982. *Laboratory Medicine Hematology*, 6th ed. pp. 116–210, 658–670, 867–869, 872. C. V. Mosby, St. Louis.

Pearse, A. G. E. 1985. *Histochemistry, Theoretical and Applied*, Vol. 2, 4th ed., pp. 686–692. Churchill Livingstone, Edinburgh.

OBSERVATIONS AND QUESTIONS

1. Examine the Wright–Giemsa–stained preparation with the low power and oil-immersion objectives. Do not use any colored filters. Why not?

_____

2. Under oil-immersion, identify a neutrophil, lymphocyte, monocyte, eosinophil, basophil (if encountered), erythrocytes, and platelets.

3. Using the high-dry and oil-immersion objectives without any colored filters, now scan the *PAS* slide, both the portion that was counterstained and the portion that was not counterstained. Do not let the oil-immersion objective touch the sticky mountant. Identify neutrophils, small and medium lymphocytes, eosinophils, and platelets. It may not be possible to distinguish large lymphocytes from monocytes. What region of the leukocytes has the Schiff's reagent stained?

_____

What region of the leukocytes has the hematoxylin stained?

_____

4. Which type of leukocyte has the highest concentration of PAS-positive material?

_____

Which type of leukocyte has the lowest concentration of PAS-positive material?

_____

Locate an eosinophil. Describe the staining response of the eosinophil granules.

_____

5. In a region of the smear that was not counterstained, identify the erythrocytes. Where is the erythrocyte stained most intensely?

_____

6. Identify the platelets and describe their staining response.

_____

7. Examine the diastase-treated control slide. How does the PAS staining in the neutrophils, lymphocytes, eosinophils, and platelets compare with that previously observed on the *PAS* slide?

_____

_____

What do your observations reveal about the amount of glycogen in each type of leukocyte?

_____

_____

8. Examine the erythrocytes in a region of the smear that was not counterstained. How does the staining compare to that observed on the PAS slide without diastase treatment?

_____

What do your observations from both slides reveal about the chemical nature of the PAS-positive material in erythrocytes?

_____

_____

9. What is an advantage of having a counterstained preparation?

_____

_____

What is an advantage of having a preparation that is *not* counterstained?

_____

_____

_____

CELLULAR NUCLEIC ACIDS

The nucleic acids are the primary informational macromolecules in the cell. There are two classes of nucleic acid, *deoxyribonucleic acid*, or *DNA*, and *ribonucleic acid*, or *RNA*. They specify the amino acid sequences of all proteins, including, of course, enzymes. Since the repertory of enzymes and structural proteins determines the cell's biochemical and physiological potential, it is ultimately the nucleic acids that are responsible for the phenotype.

Nearly all the DNA is contained within the nucleus. *Messenger RNA (mRNA)* is synthesized in the nucleus and carries the code contained in the DNA to the cytoplasm, specifically, to the *ribosomes*. The ribosomes are composed of protein and a different type of RNA, termed *ribosomal RNA (rRNA)*. The bulk of the cell's RNA is rRNA. Furthermore, a cell that is active in protein synthesis has a high concentration of cytoplasmic RNA. Also involved in the process of protein synthesis is yet a third type of RNA, *transfer RNA (tRNA)*, which transports the individual amino acids to the ribosomes and decodes the nucleic acid message contained in the mRNA. Both DNA and RNA can be identified cytochemically with the *methyl green–pyronin method*. The objectives of this project are to use the methyl green–pyronin method to localize the two classes of nucleic acid and to estimate their relative concentrations in different blood cell types.

STRUCTURE OF RNA AND DNA

The structures of RNA and DNA are shown in Figures 3.1 and 3.2. Both RNA and DNA are polymers of *nucleotides* (Fig. 3.1a), each consisting of a 5-carbon sugar (*ribose* in RNA and *2'-deoxyribose* in DNA) covalently bonded to a phosphate at the 5' end and a nitrogenous base (Fig. 3.1b) at the 1' end. Also shown are the polynucleotide chains (Fig. 3.2 a, b) and the DNA *double helix* (Fig. 3.2c), in which the two chains are held together by hydrogen bonds between *complementary bases*, i.e., between *adenine* (*A*) and *thymine* (*T*) and between *guanine* (*G*) and *cytosine* (*C*). Each RNA molecule is composed of only one polynucleotide chain. It also differs from DNA in that the base *uracil* (Fig. 3.1b) replaces thymine. The single chain of RNA folds back on itself in places so that there often are both single-stranded and double-stranded regions in the same RNA molecule. The double-stranded regions are stabilized by hydrogen bonds between complementary bases, as in DNA.

METHYL GREEN–PYRONIN METHOD

The methyl green–pyronin method was introduced in the 1940s by Jean Brachet, whose cytochemical studies of the amount of RNA in animal tissues were among the earliest to suggest a link between RNA and protein synthesis. The staining procedure for the methyl green–pyronin reaction is rather straightforward: the fixed tissue, on

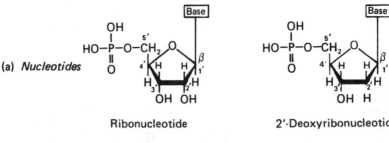

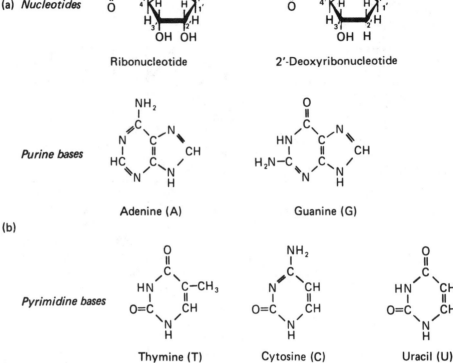

FIGURE 3.1
The building blocks of nucleic acids. The repeating unit in RNA and DNA is the nucleotide (a), which consists of a nitrogenous base, a pentose sugar, and a phosphate, all covalently bonded. In RNA, the 5-carbon sugar is ribose, while in DNA it is 2'-deoxyribose. The base is a purine or pyrimidine (b); A, G, T, and C occur in DNA, while A, G, U, and C occur in RNA. A related structure is the *nucleoside*, which consists only of the sugar and base, lacking the phosphate.

slides, is immersed in a mixture of the two dyes, after which the intensely stained tissue is partially destained with a water rinse and an alcohol solution. Methyl green has a preferential affinity for DNA, which stains green, and pyronin for RNA, which stains rose-red. Because of the specificity of the staining and the relative ease with which the procedure is carried out, the methyl green–pyronin method remains a useful tool for demonstrating cellular nucleic acids, especially for estimating the RNA content of different tissues.

The precise mechanism for the staining specificity of the methyl green–pyronin reaction is not known. One model has been proposed by Scott (1967), who analyzed the molecular interactions of each dye with the two nucleic acids. He found that pyronin and certain other dyes with a planar structure interact most strongly with single-stranded nucleic acids that have more freely accessible bases, as occurs in RNA. Methyl green, on the other hand, has a nonplanar structure that interacts best with nucleic acids that have an intact double helical structure, as occurs in native DNA.

In addition to its staining specificity, a useful feature of the methyl green–pyronin reaction is its stoichiometry; the higher the concentrations of DNA and RNA, the greater the intensity of the methyl green staining and pyronin staining, respectively. With a technique known as *microspectrophotometry*, the intensity of the staining can be measured quantitatively, and the amount of each nucleic acid thereby determined. The pyronin reaction is commonly used for determining the amount of RNA in cells. While methyl green staining has been used for determining the DNA content of cells, it is generally the *Feulgen reaction* (discussed in Project 4) that is used for such microspectrophotometric analyses.

RNase-Treated Control Slide. The usual way to confirm the staining specificity for RNA is to run two parallel slides: an untreated slide stained with the methyl green–pyronin mixture and a control slide pretreated with *ribonuclease* (*RNase*), an enzyme that catalyzes the degradation of RNA. The purpose of the RNase control is to

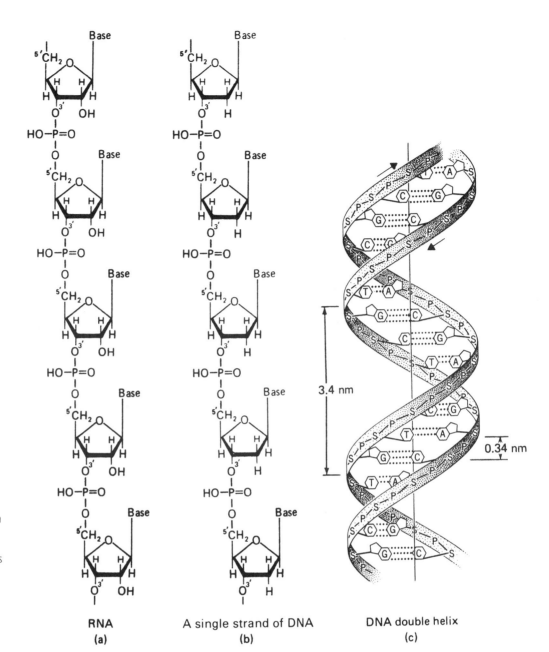

FIGURE 3.2
Structure of RNA and DNA. Each molecule of RNA (a) is a single polynucleotide chain. Each strand in DNA has a similar structure (b), but native DNA consists of two polynucleotide chains in the form of a double helix (c). The two chains are held together by hydrogen bonds (dotted lines) between A and T and between G and C.

RNA
(a)

A single strand of DNA
(b)

DNA double helix
(c)

3.4 nm

0.34 nm

confirm that the structures stained with pyronin on the untreated slide do so because they contain RNA. The use of an RNase-treated control is so routine that the procedure is sometimes referred to as the *ribonuclease, methyl green–pyronin method* (Pearse, 1968).

Slide Preparation.   The staining protocol selected for this project was developed by Perry and Reynolds (1956). You will be provided with two blood smears that have been fixed in *Carnoy's fluid*, a 6:3:1 mixture of ethanol:chloroform:acetic acid. This fixative preserves nucleic acids and keeps the white blood cells affixed to the slide during the procedure. (Red blood cells hemolyze and are not visible.) As in Project 2, all the steps are carried out in Coplin jars. As the slides are being processed, the slide surface with the cell preparation must never be allowed to rest against another slide.

Following treatment of one fixed blood smear with RNase, both slides are placed in the methyl green–pyronin staining solution. After staining, it is necessary to remove excess stain, a process termed *differentiation*. Each slide is rinsed briefly in distilled water and then placed in a 3:1 mixture of *tert*-butanol:ethanol, as described by E. B. Taft (see Clark, 1973). The rinse in the alcohol solution dehydrates the preparation so that the slides can be transferred directly to xylene. After a second xylene rinse, a coverslip is mounted (see Project 2, p. 19).

PROCEDURES

METHYL GREEN–PYRONIN METHOD

1. Label the slides in pencil with your initials and the staining procedure: one slide, labeled *MGP*, will simply be stained with the double-dye mixture; the other slide, labeled *RNase + MGP*, will be pretreated with ribonuclease. The pencil markings will also serve to identify the slide surface with the cell preparation.

2. Place the slide labeled *RNase + MGP* in a 0.1% aqueous solution of RNase, pH 6.5–7.0, at 37° C, for 30 min. The enzyme treatment and all staining procedures are done in Coplin jars; use a zigzag arrangement of up to nine slides per jar. Always use forceps to place slides in or to remove slides from Coplin jars. During the enzyme treatment, the slide labeled *MGP* can simply be left on the lab bench.

3. Rinse the RNase-treated slides under running tap water for 2 min, in a Coplin jar. Then replace with distilled water and pour it off.

4. Place both slides (*MGP* and *RNase + MGP*) in the methyl green–pyronin staining solution for 30 min.

5. Rinse each slide under running distilled water for 2 to 3 sec. Then quickly drain each slide, wipe the back of each with a Kimwipe, and allow the slides to air-dry in a near-vertical position.

6. Place the slides in the butanol-ethanol solution (in a fume hood) for 2 min.

7. Drain each slide, wipe the back of each, and place in xylene (in a fume hood) for 2 min, agitating each slide at the outset. Transfer to a second xylene rinse (in a fume hood) for 2 min.

8. Remove each slide from the second xylene rinse. Before all the xylene has evaporated from the slide surface, add a drop of Permount and a coverslip. Position the coverslip so that it will cover all the smear. Lower the coverslip slowly to minimize air bubbles and then gently press out the excess liquid with a paper towel.

REFERENCES

Clark, G. 1973. *Staining Procedures*, 3rd ed., p. 155. Williams & Wilkins, Baltimore.

Pearse, A. G. E. 1968. *Histochemistry, Theoretical and Applied*, Vol. I, 3rd ed., pp 268–272. Little, Brown, Boston.

Perry, S. and Reynolds, J. 1956. Methyl–green–pyronin as a differential nucleic acid stain for peripheral blood smears. *Blood 11*:1132–1139.

Scott, J. E. 1967. On the mechanism of the methyl green–pyronin stain for nucleic acids. *Histochemie 9*:30–47.

OBSERVATIONS AND QUESTIONS

1. Using the high-dry and oil-immersion objectives without any colored filters, scan the slide that shows the methyl green–pyronin staining. Do not let the oil-immersion objective touch the sticky mountant. Identify the different types of leukocytes: neutrophils, small and medium lymphocytes, and eosinophils. It may not be possible to distinguish large lymphocytes from monocytes. If necessary, refer to Figure 2.1, page 17.
Which type of leukocyte has the most intense pyronin staining?

   _____

   Which type of leukocyte has the least intense pyronin staining?

   _____

2. Examine the RNase-treated slide and describe the pyronin staining in the leukocytes.

   _____

   What do your observations indicate about the location of RNA?

   _____

3. On the basis of staining intensity on the *MGP* slide, which type of leukocyte has

   the highest concentration of RNA? _____
   From what you know about the functions of the various types of white blood cells, give a reasonable explanation for the observed high concentration of RNA in these cells.

   _____

   _____

   _____

   Which type of leukocyte has the lowest concentration of RNA?

   _____

4. Compare the color of the nuclei in lymphocytes on the *MGP* and RNase-treated slides. How do they differ?

   _____

   _____

   Explain the difference.

   _____

   _____

   _____

5. On the RNase-treated slide, compare the intensity of the methyl green staining in small lymphocytes and large lymphocytes (or monocytes). How do they differ?

_____

Explain the difference.

_____

_____

_____

6. Suggest a method that would demonstrate the staining specificity of methyl green.

_____

_____

STAINING OF CHROMOSOMAL DNA

INTRODUCTION

The chromosomes, as the repositories of the cell's genetic information, have long been the subject of intensive research. Since they are one of the largest structures in the cell, they were among the first cell components to be studied with the light microscope. Chromosomes were so named (by W. Waldeyer, around 1890) because of their tendency to stain intensely with many dyes (*chromo* = color; *soma* = body). They can also be identified with cytochemical reactions that are specific for DNA. The *Feulgen reaction* is one of the most important cytochemical reactions for DNA and is also an excellent stain for chromosomes. An objective of this project is to carry out the Feulgen reaction on root tips of the broad bean, *Vicia faba*. The preparations are ideal for studying chromosome morphology, the other objective of the investigation.

THE CELL CYCLE

Cells in mitotically active tissues are alternately in *mitosis* and *interphase* as they proceed through the *cell cycle* (Fig. 4.1). It is during interphase that cell growth and DNA synthesis occur, the latter during the S period. There are two other periods in interphase: G_1 (gap 1), which precedes DNA synthesis, and G_2 (gap 2), which follows it. In *Vicia faba*, the total duration of the cell cycle is 19.3 hr, with mitosis occupying 2.0 hr (Evans and Scott, 1964). Since mitosis lasts approximately 10% of the duration of the cell cycle, approximately 10% of the meristematic cells are in mitosis.

CHROMOSOME MORPHOLOGY

The chromosomes of *V. faba* are a favorite subject of study by cytogeneticists. In this species, the chromosomes are very large and the diploid number is relatively low ($2n = 12$). Thus, the details of chromosome morphology can readily be observed.

The Interphase Nucleus. In the interphase nucleus, the chromosomal material, or *chromatin*, is dispersed as a rather homogeneous mass of fibers. There are usually several small clumps of condensed, deeply stained *heterochromatin*, and there is at least one *nucleolus* (pl., *nucleoli*). Nucleoli are rich in RNA and protein, as they are the structures in which the ribosomal subunits are assembled. Each nucleolus appears as a large, round or oval structure within the nucleus. As interphase cells proceed into mitosis, the nucleolar material disperses and the chromatin condenses into compact structures, the chromosomes.

Metaphase Chromosomes. The most useful stages for the study of chromosome

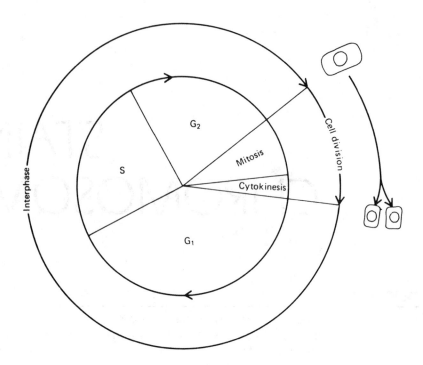

FIGURE 4.1
The cell cycle.

morphology are late prophase and early metaphase. During these stages, the chromosomes are condensed but not yet aligned on the metaphase plate. Although the chromosomes may be more condensed at metaphase, there is usually more overlapping of chromosomes when they are lined up on the spindle. The chromosomes remain condensed until telophase, when they start to decondense prior to entering the next interphase.

Metaphase chromosomes have their characteristic morphologies, as depicted in Figure 4.2. The *centromere*, also termed the *primary constriction*, is the narrow region where the two chromosomal subunits, or *chromatids*, are joined. *Metacentric* chromosomes have a median centromere and two arms of equal length. In *submetacentrics*, the centromere is closer to one end; in *acrocentrics*, the centromere is very close to the end; and in *telocentrics*, it is at the tip. In *V. faba* there are five pairs of acrocentric chromosomes and one pair of metacentric chromosomes; each metacentric is about twice as long as each acrocentric.

Another morphological feature of certain chromosomes is the *secondary constriction*, shown in the chromosome at the far right in Figure 4.2. The secondary

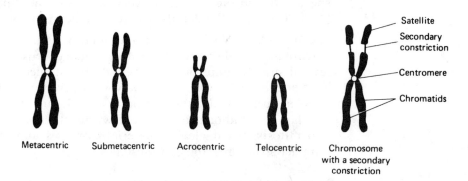

FIGURE 4.2
Morphology of metaphase chromosomes.

constriction is the very thin, lightly stained chromosomal region where a nucleolus forms at telophase. It is also the site where ribosomal RNA is transcribed. The chromosomal region distal to the secondary constriction is termed a *satellite*. In *V. faba* there is a secondary constriction in one pair of chromosomes.

FEULGEN REACTION

The Feulgen reaction was introduced by R. Feulgen and H. Rossenbeck in 1924. The method gained widespread use in cell research because it is highly specific for DNA and also permits the DNA content of individual cells to be quantified by microspectrophotometry. As noted in Project 3 (p. 24), this technique is used to measure the intensity of a dye that binds stoichiometrically to some class of molecule in the cell. With the Feulgen reaction, the DNA content of a cell can be quantified. Such measurements permit the relative amounts of DNA to be determined at different stages of mitosis and meiosis.

The Feulgen reaction will be carried out on root tips that have previously been fixed in 1:3 acetic acid:ethanol and stored in 70% ethanol. The chemical reactions that take place during the Feulgen reaction are summarized in Figure 4.3. The procedure utilizes the same dye (Schiff's reagent) used to stain carbohydrates in the PAS reaction (Project 2), but in the Feulgen reaction the tissue is treated with hot hydrochloric acid (HCl) instead of with periodic acid. HCl hydrolysis removes the purines from DNA, leaving an exposed aldehyde group at the 1' position of the depurinated deoxyriboses. The tissue is then treated with Schiff's reagent, a colorless

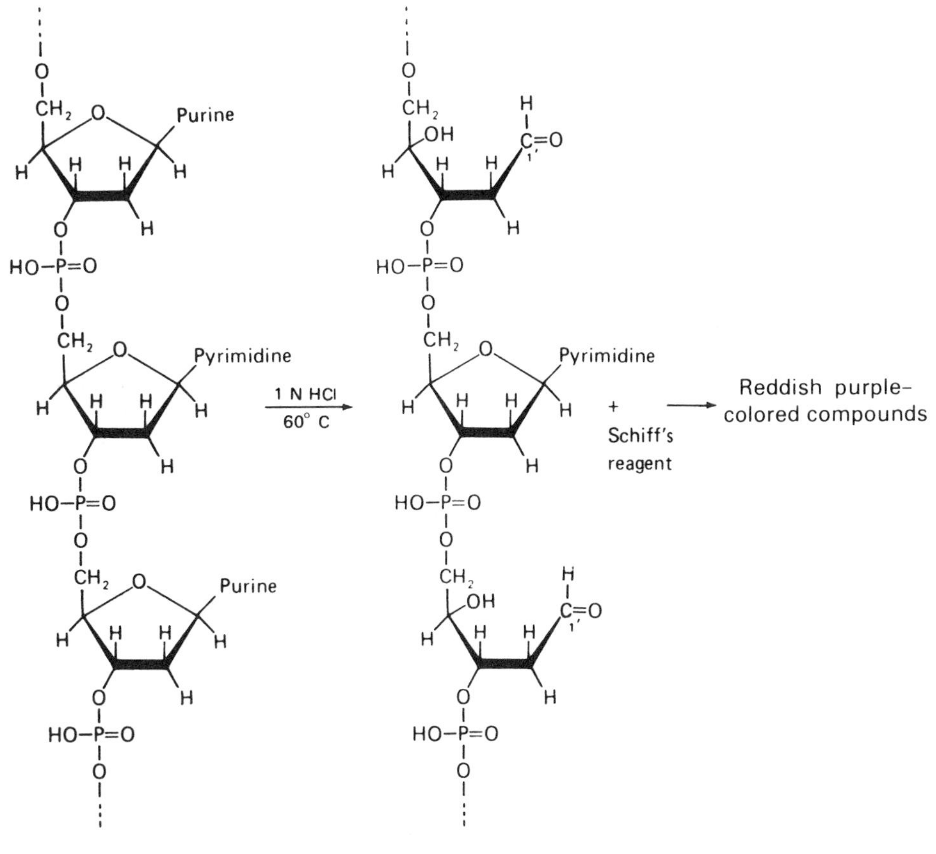

FIGURE 4.3
Steps in the Feulgen reaction.

liquid prepared by treating basic fuchsin (a red dye) with sulfurous acid. The decolorized derivative of the basic fuchsin reacts with the aldehydes on the depurinated DNA (Fig. 4.3) to form a reddish purple product. Research by Scott and Harbinson (1971) suggests that each molecule of Schiff's reagent actually combines with *two* aldehyde groups. Ribonucleic acid is not stained by the Feulgen reaction since all cellular RNA is removed by the HCl hydrolysis.

After treatment with Schiff's reagent, the tissue is rinsed with a bisulfite bleaching solution. This rinse ensures that the purple coloration is the result of Schiff's reagent having reacted with aldehydes rather than having reverted to fuchsin during the staining procedure. After the bisulfite rinse, the root tips are ready for squashing.

Trichloroacetic Acid-Treated Control Slide.   Various controls are sometimes run concomitantly with the Feulgen reaction. In this project, a root tip pretreated with hot *trichloroacetic acid* (*TCA*) will serve as a control. The hot TCA extracts *all* cellular nucleic acids, i.e., both DNA and RNA. Since the observations to be made on the TCA-treated preparation are brief, only one such control slide need be prepared by each team. The TCA-treated preparation will be counterstained.

Counterstaining.   Because of the specificity of the Feulgen reaction for DNA, only the chromosomes and interphase nuclei stain. They stain reddish purple, while the cytoplasm remains unstained. It is possible to observe the cell margins by closing the substage condenser iris of the microscope. Alternatively, a counterstain can be used to stain the cytoplasm. If the slides are to be made permanent, the cells can be counterstained with *fast green FCF*, which stains the cytoplasm bright green. In such counterstained preparations, the Feulgen staining appears violet. Since the Feulgen staining is different in a counterstained preparation, most observations will be made on a non-counterstained preparation.

SQUASH PREPARATIONS

In the root tip preparation, it is desirable to squash only the apical meristem, for it is here that the dividing cells are found. Accordingly, the nonmeristematic tissue above and below the apical meristem should be removed prior to squashing. The root cap, a protective layer of nondividing cells at the tip of each lateral root, is removed by gently rubbing the very tip of the root on a paper towel. The nonmeristematic tissue above the apical meristem is removed by cutting away all but the terminal 1.5 mm of the root tip.

The squash preparation is invaluable for the study of both plant and animal chromosomes. Unlike sectioned material, in which only a part of the cell may be present, each cell in a squash preparation contains *all* the chromosomes. The squashing also flattens the cells so that the chromosomes are more dispersed and are often in the same focal plane.

To make a flat squash preparation, the tissue must first be softened. Plant tissues, which are generally rather tough, are treated either with acid or with enzymes that digest substances in the cell wall. The Feulgen reaction is therefore especially convenient for the study of plant chromosomes since much of this softening is effected by the HCl treatment. Further softening is also facilitated by the squashing fluid, 45% acetic acid. The root tip is placed in a drop of 45% acetic acid on a slide, and the tissue is broken up by tapping it with the end of a glass rod. The acetic acid provides a medium of ideal viscosity for dispersing cells. After a coverslip is placed on the preparation, the cells are squashed by application of thumb pressure on the coverslip. A rolling motion of the thumb is used with a paper towel between the thumb and coverslip, so that the liquid pressed to the edges of the coverslip will be absorbed. For the relatively tough cells of the root tip, maximum thumb pressure is used, and a final round of squashing is done with the eraser end of a pencil.

PERMANENT, COUNTERSTAINED PREPARATIONS

The squash preparation can be made permanent with the *quick-freeze method* of A. D. Conger and L. M. Fairchild. In this method, the squashing liquid (45% acetic acid between the slide and the coverslip) is frozen, after which the coverslip is pried off with a single-edge razor blade. All, or nearly all, of the tissue adheres to the slide, which is then immediately immersed in ethanol. For counterstaining, the slide is then immersed briefly in a solution of fast green. After two ethanol rinses, a coverslip is mounted with a drop of *Euparal*, a synthetic resin that is compatible with ethanol.

PROCEDURES

For this project, you will be working in teams of four.

FEULGEN REACTION

Each team will be supplied with nine root tips in a small beaker of 70% ethanol. Eight root tips will be subjected only to the Feulgen reaction; one root tip will be pretreated with hot TCA.

1. Using forceps, transfer one root tip from the 70% ethanol to a small beaker of distilled water for a few minutes. Always grasp each root tip by the broad end. Remember to rinse the forceps with water after transfer of root tips to or from any caustic solution.
2. Transfer the root tip from the distilled water to a beaker of 5% TCA,* 90° C (in a fume hood), and treat for 15 min.
3. Transfer (using forceps) the TCA-treated root tip to a beaker of distilled water, in which it should remain until all root tips are ready for the HCl treatment.
4. Decant the 70% ethanol from the beaker containing the remaining (untreated) root tips and add distilled water to the beaker. After several minutes, transfer these root tips (using forceps) to a small beaker of 1 N HCl at 60° C. Also transfer the TCA-treated root tip to a separate beaker of 1 N HCl at 60° C.
5. After the root tips have been in the hot HCl for 10 min, dilute and cool the acid in each beaker by slowly adding distilled water.
6. Decant the liquid and then add about 20 ml of Schiff's reagent[†] (in a fume hood) to each beaker. Cover each beaker with Parafilm and then with aluminum foil to prevent light from entering. Treat with Schiff's reagent for 30 min.
7. In a fume hood, carefully decant the Schiff's reagent and add 20 ml of freshly prepared bisulfite bleaching solution (180 ml distilled water, 10 ml 10% Na or K metabisulfite solution, 10 ml 1 N HCl) to each beaker. Decant after 2 min and repeat this bleaching procedure two more times.
8. Rinse the root tips with several changes of distilled water. The root tips can be kept in the distilled water until the squash is made.

---

*CAUTION: *TCA is a strong acid. If any spills on your skin or clothing, wash it off immediately with water. In addition, protective goggles should be worn.*

[†]CAUTION: *Schiff's reagent causes burns; if any spills on your skin or clothing, wash it off immediately with water. Also avoid breathing vapors. Although it is a colorless liquid, it will stain skin, clothing, and other surfaces purplish red. Always keep the staining beaker on paper towels or Benchkote.*

SQUASH PREPARATION

Each team member should prepare two squash preparations of untreated root tips, one of which will be counterstained. One member of each team should also prepare a squash of the TCA-treated root tip.

1. Label (in pencil) an alcohol-cleaned slide and add a drop of 45% acetic acid to the center.
2. Pick up a root tip by the broad end, with forceps, and gently rub the tip on a piece of paper towel to remove the root cap. Immediately place the root tip in the drop of acetic acid.
3. With a single-edge razor blade, cut off and discard all but the terminal 1.5 mm, which contains the dividing cells. Allow the tissue to remain in the acetic acid for about 1 min.
4. Tap the root tissue with the flat end of a glass rod until the tissue is thoroughly dispersed in the drop of acetic acid. Then, gently lower a coverslip on the preparation.
5. Before squashing, remove the excess liquid as follows. Place a paper towel on the slide, hold the towel near one end of the slide, and then move the index finger of the other hand across the slide, pressing lightly. *Be very careful not to move the coverslip laterally*, as this will cause the cells to fold over on themselves, ruining the preparation.
6. With a paper towel held in place on the slide, squash the preparation with the thumb. Use a rolling motion so that the liquid is pressed to the edges of the coverslip where the towel can absorb it. Thumb pressure should be increased in successive rounds of squashing instead of being applied all at once. After applying maximum pressure with the thumb, carry out a final round of squashing with the eraser end of a pencil. *Again, be careful not to move the coverslip laterally.*

PERMANENT PREPARATIONS AND COUNTERSTAINING

The squash of the TCA-treated root tip should be counterstained. Only one of the other two squashes should be counterstained. All slides to be made permanent, whether counterstained or not, should be taken through steps 1 and 2. Always use forceps when placing slides in or removing slides from Coplin jars. All staining materials should be kept on Benchkote or paper towels.

1. Place the slide on a flat cake of dry ice* for 5 min.
2. Pry off the coverslip with a single-edge razor blade and *immediately*, before the preparation thaws, immerse the slide in a Coplin jar of 95% ethanol for 1 min. Keep track of which slide surface contains the tissue. To counterstain the TCA-treated squash and an untreated squash, proceed to steps 3 and 4. To make a permanent preparation of the other untreated squash, proceed directly to step 4.
3. Pass the slide through a second 1-min rinse in 95% ethanol and then immerse it for 10 sec in the counterstain, a 0.5% solution of fast green in 95% ethanol.
4. Rinse the slide in two changes of absolute ethanol, 1 min each with occasional agitation, and then mount a fresh coverslip with Euparal. Lower the coversip slowly to minimize air bubbles and then press out the excess liquid with a paper towel.

*CAUTION: *Dry ice is extremely cold and can injure the skin on contact.*

STAINING OF CHROMOSOMAL DNA

REFERENCES

Evans, H. J. and Scott, D. 1964. Influence of DNA synthesis on the production of chromatid aberrations by X rays and maleic hydrazide in *Vicia faba*. *Genetics 49*:17–38.

Gahan, P. B. 1984. *Plant Histochemistry and Cytochemistry*, pp. 35–37, 197–198. Academic Press, Orlando, FL.

Pearse, A. G. E. 1985. *Histochemistry, Theoretical and Applied*, Vol. 2, 4th ed., pp. 622–630, 850–852. Churchill Livingstone, Edinburgh.

Scott, J. F. and Harbison, R. J. 1971. The Schiff reaction of polyaldehydes. *Proc. R. Microsc. Soc. 6*:22–23.

Swanson, C. P., Merz, T., and Young, W. J. 1981. *Cytogenetics*, 2nd ed., pp. 77–87, 184–193. Prentice–Hall, Englewood Cliffs, NJ.

OBSERVATIONS AND QUESTIONS

1. Using the low-power and high-dry objectives without any colored filters, examine the counterstained squashes of the untreated and TCA-treated root tips. How does the staining in the TCA-treated preparation differ from that in the untreated preparation?

_____

_____

Are these observations consistent with the stated specificity of the Feulgen reaction? Explain.

_____

_____

2. Examine the untreated root tip preparation that was *not* counterstained. All remaining observations should be made on this slide. Using the low-power and high-dry objectives, examine several cells, first without any filters and then with a green filter for improved contrast. Examine several interphase cells, noting the prominent nucleoli. What does the staining response of the nucleolus suggest about its chemical composition?

_____

How many nucleoli are there per nucleus?

_____

Describe the relationship between the number of nucleoli in a cell and the size of each nucleolus.

_____

_____

3. Locate a cell at late prophase or early metaphase where the chromosomes are well spread and condensed. Using oil-immersion, locate the acrocentric and metacentric chromosomes. Is the secondary constriction located in an acrocentric pair or in the metacentric pair?

_____

On page 39, draw a typical acrocentric chromosome and one of the metacentric chromosomes, labeling the structures observed. Using the ocular micrometer, measure these chromosomes in the slide preparation and enter the lengths (in μm) next to the chromosomes in your drawing. Be sure to include the plate magnification as well.

4. Examine the secondary constrictions in several cells. Is There Feulgen-positive material in this chromosomal region? _____ Explain.

_____

_____

5. How many secondary constrictions should be present in a cell at *anaphase*?

_____ Examine several cells at anaphase to verify your prediction. On page 39, draw a cell at anaphase, labeling each of the secondary constrictions.

6. In the Feulgen reaction, occasionally a control slide *not* treated with HCl will be processed. From your knowledge of the chemical basis for the Feulgen reaction, predict how such a cell preparation should stain.

_____

Explain.

_____

_____

What would be the purpose of such a control?

_____

_____

SPECTROPHOTOMETRY OF DNA AND RNA

INTRODUCTION

Many methodologies have been developed for extracting, characterizing, and quantifying the various classes of DNA and RNA. Nucleic acids are often identified by their distinctive *absorption spectra*. An absorption spectrum is a curve showing how much light a substance absorbs at various wavelengths. All nucleic acids absorb intensely in the ultraviolet (UV) range, a result of the conjugated double bonds in the purines and pyrimidines (see Fig. 3.1b). This project consists of two parts, both of which are spectrophotometric analyses of DNA and RNA. The objective of the first part is to obtain the absorption spectrum for pure samples of DNA and RNA using UV spectrophotometry. In the second part of the project, DNA and RNA are extracted from bovine liver. The partially degraded nucleic acids in the extract are converted to colored compounds that can be measured spectrophotometrically. The objective of the second part is to determine the concentrations of DNA and RNA in bovine liver using the above-described methodology.

SPECTROPHOTOMETRY

Spectrophotometry is a method used to measure how much radiant energy a substance absorbs at different wavelengths. Pigments and other colored materials absorb in the visible range (380–760 nm), although many substances absorb maximally at shorter wavelengths (in the UV) or at longer wavelengths (in the infrared, IR). From the absorption spectrum of an unknown substance, one can often identify it or, at least, place it in a class of compounds. Especially useful for identification is the *absorption maximum*, the wavelength at which the absorption

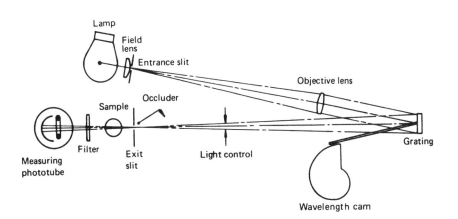

FIGURE 5.1
Optical system of a
spectrophotometer.
(Courtesy of Milton Roy.)

41

spectrum exhibits a peak. Spectrophotometry can also be used for the *quantitative* analysis of a known compound. From measurement of light absorption at an appropriate wavelength, the concentration of the compound can readily be determined. In pure form, the nucleic acids can be quantified by absorption measurements in the UV range.

Spectrophotometer Components. The instrument used to measure the absorption of a sample at different wavelengths is the spectrophotometer. The optical system of a relatively simple spectrophotometer, such as the Milton Roy *Spectronic 20*, is shown in Figure 5.1. For absorption measurements between 340 and 1000 nm (near UV, visible, and near IR), the light source is a *tungsten lamp*. In spectrophotometers capable of analyses in the UV range (200–360 nm), there is also a *deuterium lamp*. The light is focused on a *diffraction grating*, which splits the light into its component colors or wavelengths. Different wavelengths are selected by changing the position of the grating with the *cam*. The final beam that strikes the sample is relatively monochromatic; the spectral slitwidth is 20 nm in the simplest spectrophotometers and considerably narrower in more sophisticated instruments. Any light not absorbed by the sample will strike the *phototube*, which converts the transmitted light energy to an electric current. The electric signal is displayed on a meter that is calibrated both in terms of how much light the sample absorbs and how much light it transmits. In lieu of meters, some spectrophotometers have digital readouts.

Transmittance and Absorbance. The fraction of incident light transmitted by a sample is termed the *transmittance* (T). Using symbols, $T = I/I_0$, where I_0 is the intensity of the light that strikes the sample and I is the intensity of the light after it has passed through the sample. Transmittance is usually expressed as a percentage:

$$\%T = \frac{I}{I_0} \times 100$$

The *absorbance (A)* or *optical density* of the sample is a logarithmic function of T:

$$A = \log \frac{1}{T} = \log \frac{I_0}{I}$$

At 100% transmittance, $A = \log 1 = 0$. If a sample transmits 50% of the incident light, $A = \log (1/0.5) = \log 2 = 0.30$. The spectrophotometer usually has two scales, one calibrated in $\%T$ from 100 to 0 and the other in A units from 0 to ∞, though the last calibrated division is 2.0. In plotting spectral data, it is usually the absorbance that is plotted against wavelength or concentration.

Use of the Spectrophotometer. Prior to reading the absorbance of a sample, it is necessary to adjust the spectrophotometer for a reference "blank," which contains all the substances in the sample *except* the material being measured. Since the material being measured is dissolved in a solvent, it is necessary to compensate for the absorbance of the solvent itself. It is also necessary to compensate for the loss of light caused by reflections from the surface of the *cuvette*, the special glass (quartz for UV analyses) vial that holds the liquid sample. In most spectrophotometers there is a *single beam* of light and a sample holder that accommodates one cuvette. Prior to measuring the absorbance of the sample, the blank is inserted and the spectrophotometer adjusted so that the scale reads 100%T ($A = 0$). Any change in absorbance when the sample is placed in the light path must be due to the material being measured. If additional samples are to be read *at the same wavelength*, it is not necessary to adjust for the blank each time. When the wavelength is changed, however, the instrument must be readjusted for the blank, since the light source does not emit all wavelengths at the same intensity and the phototube is not equally

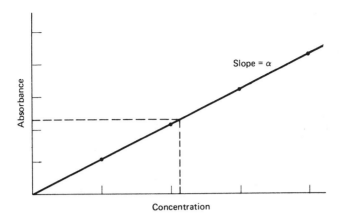

FIGURE 5.2
A typical standard curve, which is a plot of absorbance versus known concentrations of the pure substance.

sensitive to all wavelengths. The more sophisticated spectrophotometers have a *double beam*; both the sample and blank are accommodated in the sample holder, and the correction for the blank is made automatically at each wavelength. Some double-beam spectrophotometers can also scan a range of wavelengths automatically and make a permanent record of the absorption spectrum on a strip-chart recorder.

Beer–Lambert Law. For many compounds, an important relationship exists between absorbance and two sample parameters. The relationship, known as the *Beer–Lambert law*, states that absorbance is directly proportional to *solute concentration (c)* and to the *length of the light path (l)*. The equation for the relationship is:

$$A = \alpha cl$$

where the constant, α, is the *absorption coefficient*. In the biological literature, the concentration and α are usually expressed on a weight basis. In this project, c is expressed in micrograms per milliliter (μg/ml), l in centimeters, and α in centimeters squared per microgram (cm^2/μg). Since the light path (interior diameter of the cuvette) is the same for all solutions, a plot of absorbance versus concentration gives a straight line with slope α. Such a plot using *known* concentrations of the pure compound is termed a *standard curve* (Fig. 5.2). As can be seen in Figure 5.2, the standard curve can be used to determine the concentration of the same compound in a solution of *unknown* concentration. The absorbance of the unknown solution is measured and the concentration read from the graph (Fig. 5.2, dashed lines). If the Beer-Lambert equation is written as

$$c = \frac{A}{\alpha l}$$

it is clear that concentration can also be determined just from the absorbance of the unknown solution and the value of α. The absorption coefficients for biological molecules can be determined experimentally or can be found in the literature. The length of the light path, l, is usually 1 cm.

EXTRACTION OF DNA AND RNA

For the present extraction of DNA and RNA from bovine liver, you will use a quick and simple method developed by W. C. Schneider (1957). The procedure consists of tissue *homogenization*, i.e., cell disruption, followed by extractions with trichloroacetic acid (TCA) and ethanol. The first treatment is with *cold* TCA, which precipitates nucleic acids, proteins, and lipids. The precipitate is separated from the acid-soluble

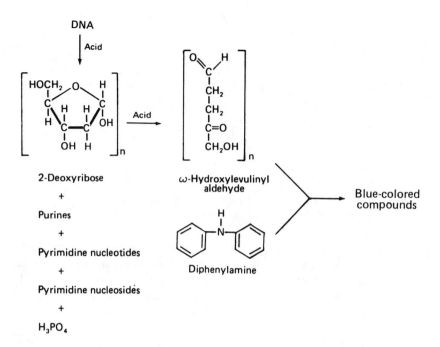

FIGURE 5.3
Diphenylamine reaction for DNA.

components by centrifugation; the *pellet*, which contains the unpurified nucleic acids, is retained, and the overlying liquid, or *supernatant*, is discarded. The precipitate is further purified with ethanol, which dissolves out the lipids. The final step is treatment with *hot* TCA, which dissolves out and partially degrades the DNA and RNA. There is some depurination and some cleavage of phosphodiester bonds.

COLORIMETRIC ASSAYS

The assay procedures are based on color reactions with the pentose moieties in the extracted nucleic acids. The concentrations of the sugars are determined by *colorimetry*, a procedure in which the concentration of a compound is measured by the intensity of the color produced when specific test reagents are added. In such colorimetric reactions, the intensity of the color, which is measured spectrophotometrically, is directly proportional to concentration over a limited range. Since the Beer–Lambert law applies, a standard curve can be constructed and the concentration of an unknown determined. The colorimetric assays you will use for DNA and RNA are, respectively, the *diphenylamine* and *orcinol reactions*.

Diphenylamine Reaction.   The diphenylamine reaction, which is depicted in Figure 5.3, is specific for 2-deoxypentoses. The reagent contains acetic and sulfuric acids. When heated with DNA, the acids cleave some of the phosphodiester bonds and hydrolyze the glycosidic linkages between the sugars and purines. In addition, the 2-deoxyribose is converted to ω-*hydroxylevulinyl aldehyde*, which reacts with diphenylamine to produce blue-colored compounds. The intensity of the blue color, which is measured at 600 nm, is directly proportional to the concentration of the sugar. You will carry out the diphenylamine reaction on four DNA solutions of known concentration, measure the absorbance of each, and plot a standard curve. By measuring the absorbance of the extract, you can obtain the DNA concentration from the standard curve.

Orcinol Reaction.   The orcinol reaction is depicted in Figure 5.4. Here, the reagent contains HCl. When heated with RNA, the acid cleaves some of the phosphodiester bonds and hydrolyzes the glycosidic linkages between the sugars and purines. The hot acid also converts the ribose to *furfural*, which, in the presence of ferric ions, reacts

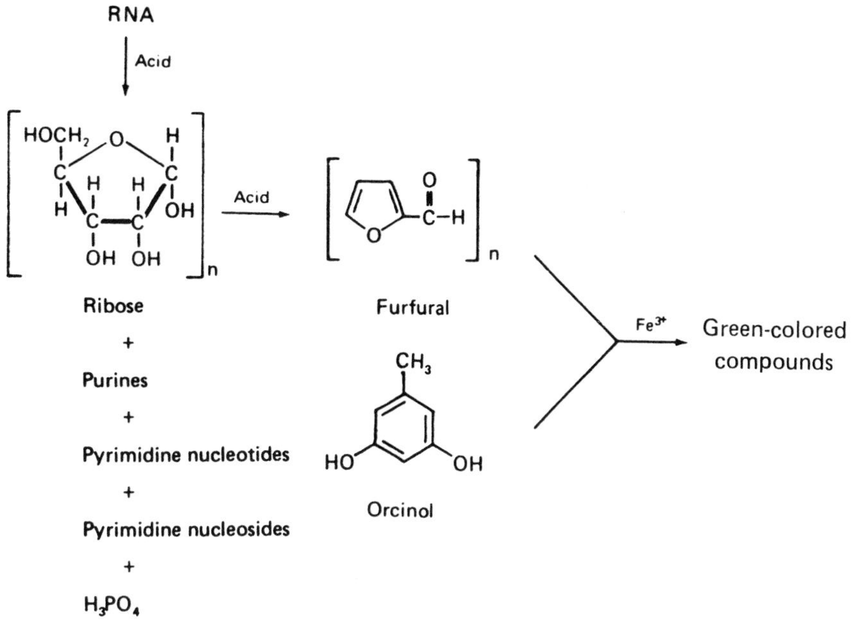

FIGURE 5.4
Orcinol reaction for RNA.

with orcinol to produce green-colored compounds. The orcinol reaction is not as specific as the diphenylamine reaction since *all* pentoses, including the deoxyribose of DNA, will react to some extent. The color contribution of DNA in the orcinol reaction is only about 10% the intensity of the same concentration of RNA. We can determine the color contribution that is due just to the RNA in the extract by subtracting the color contribution of the known concentration of DNA.

The standard curve for the orcinol reaction is generated from the absorbance readings, taken at 660 nm, for four RNA solutions of known concentration. The RNA in the standards will be from bovine liver since the RNA from each species and tissue has a characteristic purine:pyrimidine ratio. The standard curve for the orcinol reaction cannot be used directly to determine the RNA concentration in the extract since both RNA and DNA contribute to the absorbance of the extract. The absorbance due to DNA is easily estimated once the concentration of DNA in the extract has been determined; it is about 10% the absorbance of the same concentration of RNA on the standard curve. Once you have the absorbance due to DNA, you can subtract it from the absorbance reading for the extract. Using the remaining absorbance, you can determine the concentration of RNA from the standard curve for the orcinol reaction.

PROCEDURES

For all procedures in this project, you should work in pairs. The general procedures given here assume the use of single-beam spectrophotometers. For a double-beam spectrophotometer, follow the procedure in the instruction manual provided with the instrument.

UV SPECTROPHOTOMETRY

1. Turn on the power switch, select the correct lamp source (deuterium lamp), and allow the spectrophotometer to warm up for at least 5 min.
2. Add the DNA and RNA stock solutions to separate quartz cuvettes. For the reference blank, add the saline–sodium citrate solution to a third quartz cuvette; this is the solution in which the DNA and RNA are dissolved.

3. Set the wavelength at 200 nm.

4. If your spectrophotometer requires a *dark current* adjustment, proceed as follows. With the light path blocked and the cover of the sample holder closed, adjust the dark current control knob so that the needle on the scale reads 0%T.

5. Insert the reference blank in the sample holder, close the cover, and adjust the absorbance so that the needle on the scale reads 100%T ($A = 0$). When inserting any cuvette, always check that (a) the index line or other marking on the cuvette is aligned with the center marking on the sample holder, (b) the cuvette is at least half full and the liquid free of bubbles, and (c) the surface of the cuvette is free of dirt and fingerprints.

6. Replace the blank with the DNA sample, close the cover, and read the absorbance from the scale. Record all absorbance readings on Data Sheet 5.1. Continue taking absorbance readings for the DNA sample at 10-nm intervals until you have measured the absorbance at 300 nm. Remember to correct for the reference blank at each wavelength.

7. Now measure the absorbance of the RNA sample from 200–300 nm at 10-nm intervals. Remember to correct for the reference blank at each wavelength. Record the absorbance readings on Data Sheet 5.1.

NUCLEIC ACID EXTRACTION

Since enough material for the entire class is obtained from one liver sample, the homogenization should be done by one designated team. The extraction procedure is carried out by each team.

Homogenization.

1. Weigh out 30 g of frozen calf liver, cut into 1- to 2-cm cubes.
2. To a chilled Waring blender jar, add 120 ml ice-cold distilled water.
3. Turn on the blender to the highest speed and then add the pieces of almost defrosted liver, one at a time, a few seconds apart. After all the pieces have been added, continue homogenizing for 1 min.
4. Dispense 2.0 ml of the liver homogenate* to each team, adding each aliquot to a 15-ml centrifuge tube.

Extractions with TCA and Ethanol.

1. Add 5.0 ml of ice-cold 10% TCA† to the centrifuge tube. Mix well by drawing the mixture in and out of a Pasteur pipet (fitted with a rubber bulb).
2. Place the tube in a clinical centrifuge and spin at the highest speed (approximately 1300 *g*) for 2 min.
3. With the Pasteur pipet, remove and discard the supernatant. If there is a layer of solid material floating on top of the supernatant, save it by inserting the Pasteur pipet beneath this layer when the supernatant is aspirated.
4. Add another 5.0 ml of ice-cold 10% TCA to the pellet and resuspend with the Pasteur pipet. Centrifuge and aspirate as described in steps 2 and 3, leaving one or two drops of liquid above the pellet.
5. With the Pasteur pipet, disperse the pellet in the liquid above it. Then, using a graduated cylinder, add 10 ml of 95% ethanol (room temperature) to the dispersed pellet. The clumps that form should be dispersed as much as possible with the Pasteur pipet.

*CAUTION: *Do not pipet by mouth.*
†CAUTION: *TCA is a strong acid; do not pipet by mouth. If any spills on your skin or clothing, wash it off immediately with water. In addition, protective goggles should be worn.*

6. Centrifuge at 1300 g for 2 min. Decant and discard the supernatant.

7. Wash the pellet in another 10 ml of 95% ethanol, as in step 5, and then centrifuge and decant as in step 6.

8. Add 5.0 ml of 5% TCA (room temperature) to the centrifuge tube and disperse the pellet with the Pasteur pipet. Place the tube in a water bath at 90° C for 15 min, agitating the tube every few min.

9. Centrifuge at 1300 g for 2 min and then carefully decant the supernatant, which contains the degraded nucleic acids, into a test tube.

10. To make sure all the remaining solubilized nucleic acid is removed, wash the pellet with 5.0 ml of 5% TCA (room temperature), centrifuge at 1300 g for 2 min, and then combine the supernatant with the nucleic acid extract in the test tube. This combined *nucleic acid extract* will be used for the colorimetric assays.

DIPHENYLAMINE REACTION

1. Label six test tubes as shown in the following table and add the indicated volume of each solution.* Tubes 1–4 are the standards; tube 5 contains the extract with an unknown concentration of DNA. The DNA stock solutions have 5% TCA as the diluent.

Tube	[DNA] (μg/ml) in each standard	Volume of each DNA stock solution*	Nucleic acid extract*	5% TCA*
Blank	0	—	—	2.0 ml
1	100	2.0 ml (100 μg/ml)	—	—
2	200	2.0 ml (200 μg/ml)	—	—
3	300	2.0 ml (300 μg/ml)	—	—
4	400	2.0 ml (400 μg/ml)	—	—
5	—	—	2.0 ml	—

*CAUTION: *TCA is a strong acid; do not pipet by mouth.*

2. To each tube, add 4.0 ml of the diphenylamine reagent[†] dispensed from a buret. Cover each tube with Parafilm and invert three times to mix the contents.

3. Remove the Parafilm, place loose-fitting screw caps on the tubes, and then place the tubes in a beaker of vigorously boiling water (with boiling chips) for 10 min.

4. While the tubes are in the boiling-water bath, turn on the spectrophotometer; it should be allowed to warm up for at least 5 min.

5. After the tubes have been in the boiling-water bath for 10 min, transfer them to an ice-water bath to cool them quickly.

6. Carefully transfer the liquid in the six test tubes to six labeled cuvettes. The cuvettes should be labeled near the top edge with a permanent marking pen.

7. Set the wavelength on the spectrophotometer at 600 nm.

8. With no cuvette in the light path and the cover of the sample holder closed, adjust the *dark current* control knob (termed "Power Switch/Zero Control" on the Spectronic 20) so that the needle on the scale reads 0%T.

9. Insert the reference blank in the sample holder, close the cover, and adjust the absorbance (with the "100%T Control" on the Spectronic 20) so that the needle on the scale reads 100%T ($A = 0$). When inserting any cuvette, always check that (a) the index line or other marking on the cuvette is aligned with the center marking on the sample holder, (b) the cuvette is at least half full and the liquid free of bubbles, and (c) the surface of the cuvette is free of dirt and fingerprints.

[†]**CAUTION:** *The diphenylamine reagent contains concentrated acetic acid and sulfuric acid. If any spills on your skin or clothing, wash it off immediately with water. In addition, protective goggles should be worn.*

10. Read the absorbance of each standard (tubes 1–4) and of the extract (tube 5). Since the readings are taken at the same wavelength, the spectrophotometer need not be adjusted for the blank between readings. Enter the absorbance readings on Data Sheet 5.2.

ORCINOL REACTION

1. Label six test tubes as shown in the following table and add the indicated volume of each solution.* Tubes 1–4 are the standards; tube 5 contains the extract with an unknown concentration of RNA. The RNA stock solutions have 5% TCA as the diluent.

Tube	[RNA] (μg/ml) in each standard	Volume of each RNA stock solution*	Nucleic acid extract*	5% TCA*
Blank	0	—	—	3.0 ml
1	100	0.4 ml (100 μg/ml)	—	2.6 ml
2	200	0.4 ml (200 μg/ml)	—	2.6 ml
3	300	0.4 ml (300 μg/ml)	—	2.6 ml
4	400	0.4 ml (400 μg/ml)	—	2.6 ml
5	—	—	0.4 ml	2.6 ml

*CAUTION: *TCA is a strong acid; do not pipet by mouth.*

The RNA concentrations for tubes 1–4 are taken to be those listed in the table; the addition of TCA is a required dilution in the orcinol reaction, which is an extremely sensitive colorimetric reaction.

2. To each tube, add 3.0 ml of the orcinol reagent[†] dispensed from a buret. Cover each tube with Parafilm and invert three times to mix the contents.

3. Remove the Parafilm, place loose-fitting screw caps on the tubes, and then place the tubes in a beaker of vigorously boiling water (with boiling chips) for 20 min.

4. After the tubes have been in the boiling-water bath for 20 min, transfer them to an ice-water bath to cool them quickly.

5. Carefully transfer the liquid in the six test tubes to six labeled cuvettes.

6. Set the wavelength on the spectrophotometer (warmed up) at 660 nm, adjust for the blank, and read the absorbance of each standard (tubes 1–4) and of the extract (tube 5). Enter the absorbance readings on Data Sheet 5.2.

REFERENCES

Adams, R. L. P., Knowler, J. T., and Leader, D. P. 1986. *The Biochemistry of the Nucleic Acids*, 10th ed. Chapman and Hall, London.

Schneider, W. C. 1957. Determination of nucleic acids in tissues by pentose analysis. In *Methods in Enzymology*, Vol. 3, Colowick, S. P. and Kaplan, N. O., eds., pp. 680–684. Academic Press, New York.

Segel, I. H. 1976. *Biochemical Calculations*, 2nd ed., pp. 324–346. John Wiley, New York.

Slayter, E. M. 1970. *Optical Methods in Biology*, pp. 528–551. John Wiley, New York.

[†]**CAUTION:** *The orcinol reagent contains concentrated hydrochloric acid. If any spills on your skin or clothing, wash it off immediately with water. In addition, protective goggles should be worn.*

EXERCISES AND QUESTIONS

1. On Graphs 5.1 and 5.2, plot the UV absorption spectrum for DNA and for RNA, respectively. Plot absorbance (ordinate) versus wavelength (abscissa). Each curve should be carefully drawn with a French curve template.

2. At what wavelength is absorption a maximum for DNA? _____ for RNA?

_____

3. On Graph 5.3, plot the standard curve for the diphenylamine reaction. Plot absorbance (ordinate) versus the known DNA concentrations (μg/ml, abscissa) in the standards (tubes 1–4). The reference blank should also be included as a point (A = 0, [DNA] = 0). Using a ruler, draw the best-fit line for the points.

4. From your standard curve and the absorbance reading of the extract, determine the concentration of DNA in the extract. Enter the value on Data Sheet 5.2.

5. On Graph 5.4, plot the standard curve for the orcinol reaction. Plot absorbance (ordinate) versus the known RNA concentrations (μg/ml, abscissa) in the standards (tubes 1–4). The reference blank should also be included as a point (A = 0, [RNA] = 0). Using a ruler, draw the best-fit line for the points.

6. Determine the color contribution of DNA to the orcinol reaction of the extract. DNA reacts one-tenth as much as the same concentration of RNA. From the standard curve for the orcinol reaction, find the absorbance value for an RNA concentration that is the same as the concentration of *DNA* in the extract. Take one-tenth of that absorbance as the absorbance component due to DNA and enter the values.

Concentration of DNA in the extract = _____

Absorbance of RNA at concentration above = _____

Absorbance due to DNA = 0.1 (absorbance of RNA at concentration above)

= _____

7. To obtain the approximate concentration of RNA from the absorbance reading of the extract, it is necessary to subtract the absorbance component due to DNA.

Absorbance of extract _____

−Absorbance due to DNA _____

Absorbance due to RNA _____

On the standard curve for the orcinol reaction, look up the concentration of RNA for the absorbance due to the RNA. Enter the value on Data Sheet 5.2.

8. Calculate the ratio [RNA]/[DNA] in the extract using the values obtained from the standard curves. _____ What trend in this ratio would you expect to find for tissues more active in protein synthesis? Explain.

_____

_____

_____

_____

9. What is the concentration of DNA and RNA in bovine liver? _____ μg DNA/mg liver; _____ μg RNA/mg liver. Show the calculations in the space provided.

10. Assuming that the light path is 1 cm, what is the value of α for the diphenylamine reaction? _____ cm^2/μg. Show calculations.

Data Sheet 5.1
UV absorbance readings for DNA and RNA dissolved in saline–sodium citrate.

Wavelength (nm)	DNA	RNA
200		
210		
220		
230		
240		
250		
260		
270		
280		
290		
300		

Data Sheet 5.2

Absorbance readings for the diphenylamine and orcinol reactions

	Diphenylamine reaction	
Tube	[DNA] (μg/ml)	A_{600}
1	100	
2	200	
3	300	
4	400	
5 (extract)		

	Orcinol reaction	
Tube	[RNA] (μg/ml)	A_{660}
1	100	
2	200	
3	300	
4	400	
5 (extract)		

Graph 5.1

UV absorption spectrum for DNA.

Graph 5.2
UV absorption spectrum for RNA.

NAME ______________________________ SECTION __________ DATE ____________

LAB PARTNER ______________________________

Graph 5.3
Standard curve for the diphenylamine reaction.

LAB PARTNER _____

Graph 5.4
Standard curve for the orcinol reaction.

ELECTROPHORESIS OF HEMOGLOBIN

INTRODUCTION

Electrophoresis is a method commonly used for separating and characterizing charged molecules, especially proteins and nucleic acids. All electrophoretic methods are based on the principle that charged molecules will migrate through a liquid or semisolid medium that is subjected to an electric field. Each charged molecule migrates at a characteristic rate depending on its charge, size, and other physical characteristics. In the process, different charged molecules in a mixture migrate different distances and ultimately form discrete bands in the support medium. Electrophoresis is so discriminating a method that it can often separate two proteins that differ in only a single amino acid. One protein that has been studied extensively by electrophoresis is hemoglobin, the oxygen-transporting pigment in red blood cells. The objective of this project is to separate normal and abnormal forms of hemoglobin using the technique of electrophoresis with *cellulose acetate* as the support medium.

PROTEIN STRUCTURE

All proteins are polymers of amino acids. Figure 6.1 shows the general structure of amino acids and the peptide bond that links them together in proteins. Attached to the α-(alpha-) *carbon* are an α-*amino group*, an α-*carboxyl group*, a *hydrogen atom*, and a *radical* or *R group* (two different R groups, designated R_1 and R_2 are shown). The peptide bond forms between the α-carboxyl group of one amino acid and the α-amino group of another amino acid, resulting in a *dipeptide*. Chains of more than 10 amino acids are called *polypeptides*. There are 20 different amino acids commonly occurring in proteins. Each has a different R group, which not only distinguishes it from the

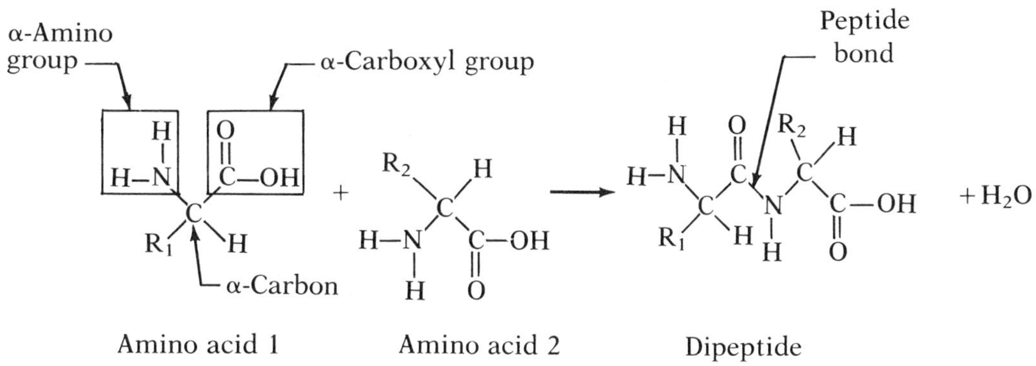

FIGURE 6.1
General structure of amino acids and formation of a dipeptide.

Amino acid 1   Amino acid 2   Dipeptide

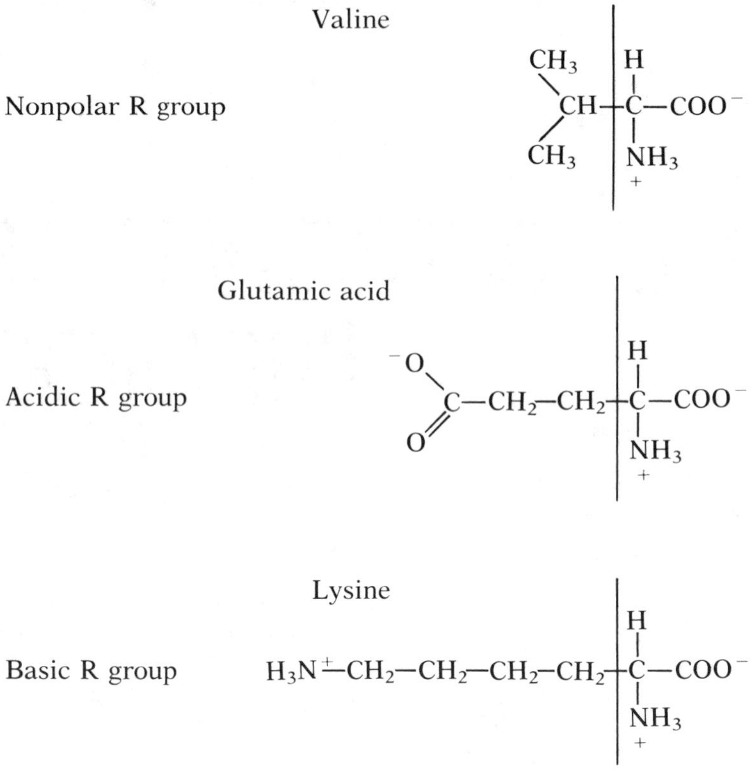

Valine

Nonpolar R group

$$\underset{CH_3}{\overset{CH_3}{\diagdown}} CH - \underset{\underset{+}{NH_3}}{\overset{H}{\underset{|}{C}}} - COO^-$$

Glutamic acid

Acidic R group

$$\overset{-O}{\underset{O}{\diagdown}} C - CH_2 - CH_2 - \underset{\underset{+}{NH_3}}{\overset{H}{\underset{|}{C}}} - COO^-$$

Lysine

Basic R group

$$H_3N^+ - CH_2 - CH_2 - CH_2 - CH_2 - \underset{\underset{+}{NH_3}}{\overset{H}{\underset{|}{C}}} - COO^-$$

FIGURE 6.2
Structures of valine, glutamic acid, and lysine.

other amino acids but also affects its chemical properties. The amino acids of particular interest in this project are depicted in Figure 6.2. They are *valine*, which has a nonpolar R group, *glutamic acid*, which has an acidic or negatively charged R group, and *lysine*, which has a basic or positively charged R group.

Some proteins are composed of one long polypeptide chain, while others contain two or more chains. The sequence of amino acids in the polypeptide is designated the *primary structure*. There are higher orders of structure that impart a three-dimensional shape, or *conformation*, to the protein. The *secondary structure* refers to the orientation of neighboring amino acids in three-dimensional space. A very common type of secondary structure is the α-helix, a right-handed coil that is stabilized by hydrogen bonds between neighboring amino acids. About 75% of the hemoglobin molecule is in an α-helix. In *globular proteins*, which include hemoglobin, other transport molecules, and enzymes, helical regions are separated by nonhelical

FIGURE 6.3
The quaternary structure of normal adult hemoglobin, which consists of four polypeptide chains, each associated with a heme group. There are two identical α chains and two identical β chains. Also depicted is a stretch of alpha helix in the β_1 chain. The letters N and C denote, respectively, the α-amino end and α-carboxyl end of each chain.

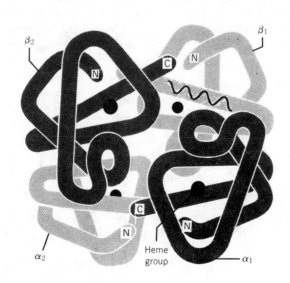

stretches of the polypeptide that twist and turn without any repeating structure. Such *random coils*, as they are termed, permit the helical regions to fold and twist, ultimately generating the *tertiary structure*, which is the conformation of the *entire* polypeptide. It is the sequence of amino acids in the polypeptide chain that determines the final conformation and function of the protein. Polar and charged amino acid residues, which are hydrophilic, remain at the surface of the molecule exposed to the aqueous environment, while nonpolar residues, which are hydrophobic, tend to cluster in the interior of the molecule. The tertiary structure is stabilized by electrostatic bonds between positively and negatively charged R groups, disulfide bridges, hydrogen bonds, and hydrophobic bonds between nonpolar residues. In the case of proteins composed of more than one polypeptide chain, the term *quaternary structure* is used to describe the orientation of the chains and the nature of the bonds that stabilize this orientation. Figure 6.3 shows the quaternary structure of normal adult hemoglobin, which contains four polypeptide chains.

HUMAN HEMOGLOBINS

There are five types of human hemoglobin that you will be separating in this project: hemoglobins A, A$_2$, F, S, and C.

Hemoglobin A.   Normal adult hemoglobin or *hemoglobin A (Hb A)* makes up 97–98% of the hemoglobin in the adult. Hb A contains two identical α-*globin* chains, each with 141 amino acids, and two identical β-*globin* chains, each with 146 residues. Thus, the formula for the Hb A tetramer can be written $\alpha_2\beta_2$. Each globin chain is associated with an iron-containing *heme group* (Fig. 6.3), a nonprotein component that is the site of oxygen binding.

Hemoglobin A$_2$.   *Hemoglobin A$_2$ (Hb A$_2$)*, a minor class (2–3%) of hemoglobin in the normal adult, contains two α chains and two δ (delta) chains, the latter having β-like sequences. The δ chain differs from the β chain in only 10 of its 146 amino acids. The formula for Hb A$_2$ is $\alpha_2\delta_2$.

Hemoglobin F.   The predominant hemoglobin from the eighth week of development until birth is *fetal hemoglobin (Hb F)*. Hb F contains two α chains and two γ (gamma) chains, the latter being analogous to the β chains. The γ chain differs from the β chain in 39 of its 146 amino acids. Fetal hemoglobin can be represented as $\alpha_2\gamma_2$. By the time a normal infant is about six months old, very little Hb F remains in the blood. Only in individuals with certain inherited blood disorders in which the β chain is not synthesized is Hb F still produced in appreciable quantities throughout adult life.

Hemoglobin S.   The first abnormal hemoglobin to be studied at the molecular level was *sickle-cell hemoglobin (Hb S)*, which is responsible for sickle-cell anemia, a serious heritable disease that occurs primarily in individuals of African ancestry. In afflicted individuals, the red blood cells have an abnormal morphology; instead of being biconcave disks, the erythrocytes are crescent- or sickle-shaped. The misshapen cells block small blood vessels, damaging internal organs, and are short-lived, resulting in anemia. At the protein level, the difference between Hb A and Hb S is but a single amino acid in each β chain. Specifically, at the sixth position from the α-amino end, glutamic acid is replaced by valine. If we designate the abnormal β chain as β^s, the Hb S tetramer can be written as $\alpha_2\beta_2^s$.

Hemoglobin C.   A much rarer variant is *hemoglobin C (Hb C)*, which is responsible for hemoglobin C disease, a less serious condition than sickle-cell anemia. The molecular basis for Hb C also involves a change at the sixth position of each β chain; in lieu of a glutamic acid, there is a lysine. The Hb C tetramer can be written as $\alpha_2\beta_2^c$.

Table 6.1 gives the tetrameric formulas of the aforementioned hemoglobins and summarizes differences in the primary structure of the β chains and differences in *isoelectric points* (discussed shortly).

Table 6.1
Structural differences and isoelectric points (pI) of hemoglobins A, A$_2$, F, S, and C.

Hemoglobin	Chain formula	Sixth amino acid in β chain	pI
Hb A	$\alpha_2\beta_2$	Glutamic acid	6.93
Hb A$_2$	$\alpha_2\delta_2$	—	7.22
Hb F	$\alpha_2\gamma_2$	—	7.00
Hb S	$\alpha_2\beta_2^s$	Valine	7.10
Hb C	$\alpha_2\beta_2^c$	Lysine	7.28

The pI values are from J. A. Koepke et al. (Reprinted with permission from *Clinical Chemistry*, 1975, Vol. 21, p. 1955, Table 1. Copyright American Association for Clinical Chemistry, Inc.)

ELECTROPHORESIS

Electrophoretic separations are possible because charged particles in a solution will migrate when subjected to an electric field. Negatively charged molecules migrate toward the positive pole, or *anode*, while positively charged molecules advance toward the negative pole, or *cathode*. In *zone electrophoresis*, the charged particles migrate through a solid or semisolid, porous support medium. For the separation of hemoglobins, you will use cellulose acetate plates as the support medium.

The basic components of an electrophoresis apparatus are shown in Figure 6.4. The two major components are the *electrophoresis chamber*, which holds the support medium and the electrolyte solution (buffer), and the *power supply*. The power supply converts the AC (alternating current) of the line source to DC (direct current), the

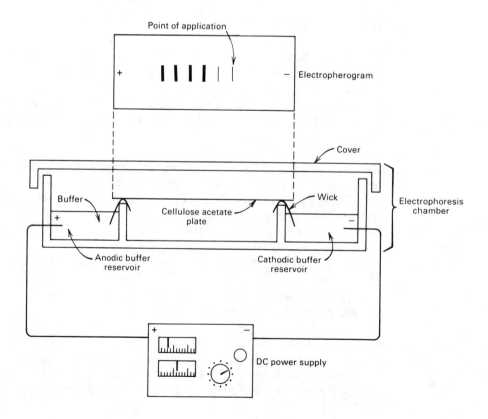

FIGURE 6.4

Components of an electrophoresis system (end view) for separating hemoglobins and, above it, the electropherogram (face view).

latter required for electrophoresis, and allows a specific voltage to be selected. The support medium containing the sample is placed between two buffer reservoirs, which are connected to the electrodes of the power supply. With certain support media, including cellulose acetate, moistened wicks serve as electrical conductors from the buffer to the plate. Since the only electrical connection between the electrodes is the buffer in the pores of the support medium, the various charged molecules in the sample migrate to the anode or cathode. If each molecular species has a unique migration rate, at the end of the process there will be a series of bands, each containing a different type of molecule. The bands are visualized using stains or staining reactions based on the biological activity of the molecules in the bands. The final pattern of bands is termed an *electropherogram* (Fig. 6.4).

Mechanism of Separation. Proteins are separated primarily due to differences in charge. The net charge depends on the number of acidic and basic amino acids in the protein *and* on the pH of the solution. Consider two proteins that differ in a single internal amino acid, say, a glutamic acid versus a lysine (see Fig. 6.2 for their structures). Since the α-amino and α-carboxyl groups of these amino acids are bonded to adjacent amino acids in the polypeptide, we need consider only the *extra* amino or carboxyl group in the radical. At a high pH (excess OH^- ions), the extra carboxyl group of glutamic acid would lose a proton and be negatively charged ($-COO^-$), whereas the extra amino group of lysine would be uncharged ($-NH_2$). While the two proteins in question may both have a net negative charge at a high pH, the one with a glutamic acid will be more negatively charged. If these two proteins differ in no other way, then during electrophoresis at a high pH, the protein with the glutamic acid will migrate to the anode more rapidly than the protein with the lysine.

For each protein there is some unique pH at which the net charge on the molecule is zero. If electrophoresis is carried out at this pH, termed the *isoelectric point* and symbolized by *pI*, the protein remains stationary. If the electrophoresis buffer has a pH above the pI of the protein, that protein will be negatively charged and migrate toward the anode, while if the pH is below the pI, it will be positively charged and migrate toward the cathode. The electrophoresis buffer used in this project has a higher pH (8.2–8.6) than any of the isoelectric points for the hemoglobins you will be studying (Table 6.1). Thus, the hemoglobin molecules will all be negatively charged and migrate toward the anode.

Let us examine the various factors that determine the rate at which a molecule migrates in an electric field. There is a *driving force* on the molecule. The driving force is the product of the net charge on the molecule and the *field strength*, defined as voltage/distance between the electrodes. Since the molecules are not migrating through a vacuum but through a buffer solution in the pores of the support medium, there will be a countering drag or *resisting force*. The resisting force is directly proportional to the size of the molecule, the viscosity of the buffer solution, and the velocity at which the molecule is moving. As electrophoresis proceeds, some final velocity is attained. It can be shown that the final velocity is directly proportional to the charge on the molecule and field strength but is inversely proportional to molecular size and viscosity of the medium.

Other factors that affect migration rate are the shape of the molecule and the nature of the support medium. As might be predicted, the less streamlined the molecule's shape, the slower will be its migration rate. Similarly, the greater the adsorption of the molecule to the matrix of the support medium, the slower will be its rate. Since there is very little adsorption to the cellulose acetate matrix, migration is not impeded, and separations can be completed in a relatively short time. Another important factor is *electroosmosis*, which refers to the tendency of the buffer to flow toward the cathode. As a result of this flow, the migration rate of negatively charged molecules is decreased, and the migration rate of positively charged molecules is increased. Molecules having no charge do not remain at the point of application but move toward the cathode, carried along by the buffer.

Electrophoretic Mobility.   A measurement occasionally used for comparing migration rates is *electrophoretic mobility*, μ (mu). Electrophoretic mobility is defined as the migration rate (cm/sec) of a particle at a field strength of 1 volt/cm. The formula used to calculate μ is

$$\mu = \frac{\text{migration rate in cm/sec}}{\text{field strength in volts/cm}}$$

Thus, the units for electrophoretic mobility are centimeters squared per second per volt ($cm^2\ sec^{-1}\ volt^{-1}$). The sign of μ is that of the net charge on the particle. Protein molecules have relatively slow migration rates, with the absolute value of μ typically between 10^{-5} and $10^{-4}\ cm^2\ sec^{-1}\ volt^{-1}$. You will be obtaining approximate values of μ for five different hemoglobins (A, A_2, F, S, and C) in three samples that contain different combinations of them. Your calculations of μ will be based on the distance traveled by each hemoglobin, the duration of the electrophoresis, the voltage applied, and the distance between the points where the cellulose acetate plate contacts the moistened wicks (the latter measurement taken to be the distance between the electrodes).

Cellulose Acetate Electrophoresis of Hemoglobin.   Cellulose acetate is a white, uniform, rather brittle material. It is supplied as plates, consisting of a thin (130 μm) cellulose acetate membrane bonded to a clear, flexible layer of *Mylar* (Du Pont), an inert plastic backing that makes the brittle cellulose acetate easier to handle. Cellulose acetate has a microporous structure much like that of a sponge, with pores several micrometers in diameter comprising about 80% of the volume. It is through these pores, which will be filled with buffer, that the compounds being separated migrate.

The electrophoresis chamber is prepared by adding buffer to the *outer* anodic and cathodic buffer reservoirs and by draping a buffer-soaked wick over the inner wall of each buffer reservoir (Fig. 6.4). Prior to applying the hemoglobin samples, the cellulose acetate plate is soaked in buffer to fill the pores of the cellulose acetate with buffer. Since each sample must be applied as a very narrow streak, you will use a specially designed applicator with tips made of very fine (and delicate!) wire. The applicator is loaded by depressing the applicator tips into sample wells containing the hemoglobin samples. The three samples used for this project are prepared mixtures of two or more different hemoglobins. *Sample 1* contains hemoglobins A and A_2; *sample 2* contains hemoglobins A, A_2, and S; and *sample 3* contains hemoglobins A, F, S, and C. Hemoglobins A_2 and C are not present in the same sample because they have approximately the same electrophoretic mobilities with the electrophoresis system you will be using. To apply the samples, the cellulose acetate plate is placed in an aligning base and the loaded applicator is set in position above it (Fig. 6.5). When the spring-loaded button on the applicator is pressed, the tips make contact with the

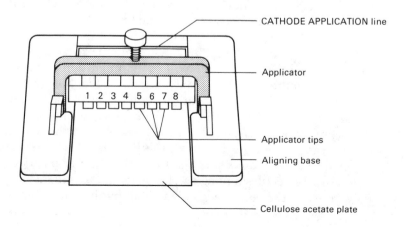

FIGURE 6.5
Application of samples to the cellulose acetate plate. See text for explanation.

plate, delivering their samples to the cellulose acetate. After a second application is superimposed on the first, the cellulose acetate plate is electrophoresed.

The plate is placed on the wicks, *cellulose acetate side down*, with the samples closer to the *cathodic* buffer reservoir. The chamber is covered and the samples electrophoresed at 350 volts for 25 minutes. *The high voltage used for electrophoresis demands that extreme caution be exercised; never touch the electrophoresis chamber or any of the electrical connections while the power is turned on.*

Following electrophoresis, the hemoglobin bands are stained with *ponceau S*, an anionic dye that, at low pH, reacts with positively charged amino groups of the protein. Background staining of the cellulose acetate is removed by soaking the plate in a dilute solution of acetic acid. Finally, the plate is dried in an oven. Before the plate is examined, it should be placed in a protective, clear plastic envelope to prevent chipping of the cellulose acetate.

In addition to the hemoglobin bands, there will also be a very faint band visible at the point of application and another faint band close to it for *carbonic anhydrase*, an enzyme in red blood cells that functions in the transport of carbon dioxide. The carbonic anhydrase is released from red blood cells during preparation of the hemoglobin sample.

PROCEDURES[1]

For all procedures, you should work in teams of four. Each team will electrophorese three samples: *sample 1* (A, A_2), *sample 2* (A, A_2, S), *sample 3* (A, F, S, C). Each team will have its own micropipetter, sample well plate, applicator, and cellulose acetate plate.

PREPARATION OF THE ELECTROPHORESIS CHAMBER

1. After making certain that the power supply is turned off and unplugged, add 100 ml of buffer* to each outer reservoir (anodic and cathodic) of the electrophoresis chamber.
2. Wearing disposable gloves, moisten two wicks with buffer. Then, drape a wet wick over the inner wall of each buffer reservoir, making certain that each wick is immersed in the buffer.
3. Cover the chamber to prevent evaporation.

APPLICATION OF THE HEMOGLOBIN SAMPLES

Care should be taken when handling the hemoglobin samples since there is always the slight possibility that they are contaminated with infectious agents. Accordingly, disposable gloves should be worn, all work surfaces covered with Benchkote or paper towels, spills wiped up, the area disinfected, and all waste materials disposed of properly.

1. Familiarize yourself with the following equipment items: the Bufferizer, a set of two containers designed to immerse the cellulose acetate plate into buffer very slowly; the micropipetter and disposable pipet tips; the sample well plate, the reservoir for the hemoglobin samples; the applicator; and the aligning base, the plastic holder that positions the cellulose acetate plate during sample application.
2. With a permanent marking pen, make a small identification mark at a corner of the cellulose acetate plate, on the Mylar (glossy) side.

*CAUTION: *The buffer is toxic; wash thoroughly after handling.*
[1]The electrophoresis procedure has been adapted from a protocol developed by Helena Laboratories and is used here with permission.

3. Place the plate in a rack and then place the rack in the lower chamber of the Bufferizer. Add buffer to the upper chamber and proceed to step 4 while the plates are soaking. The plates should soak about 5 min (after they are completely immersed in buffer) before the samples are applied.

4. Using a micropipetter, place 5.0 μl of each hemoglobin sample* (sample 1, sample 2, sample 3) in a separate well of the sample well plate. Run the separations in duplicate, adding sample 1 to wells 2 and 3, sample 2 to wells 4 and 5, and sample 3 to wells 6 and 7. The two end wells (1 and 8) should not be used since the samples on the edge of the plate tend to curve as their components separate. The easiest way to fill each well is to expel the drop at the pipet tip and then touch the drop to the well. Be sure to use a fresh pipet tip for each of the three samples. To prevent evaporation of the samples, cover the sample wells with a clean glass slide.

5. Prime the applicator tips as follows. Place the applicator in the brackets of the sample well plate and depress the tips three or four times in the hemoglobin samples. Remove the applicator from the brackets and touch the tips to a blotter. Replace the applicator in the brackets of the sample well plate and proceed quickly through the next three steps.

6. Remove the cellulose acetate plate from the soaking buffer and place it on a blotter, *cellulose acetate side up*. With another blotter, blot the cellulose acetate once firmly.

7. Place a drop of water on the aligning base and then place the plate, *cellulose acetate side up*, in the aligning base so that the top edge of the plate is on the line labeled CATHODE APPLICATION (Fig. 6.5). In this position, the samples will be applied closer to one end (cathodic end) of the plate. One person should apply the hemoglobin samples for each team.

8. Apply the hemoglobin samples as follows, keeping in mind that once the applicator tips are loaded, the samples must be applied within 15 sec. Load the applicator tips by depressing them into the hemoglobin samples three or four times. Immediately place the applicator in the brackets of the aligning base, which contains the cellulose acetate plate, depress the tips, and hold for 5 sec. Reload the applicator tips and make a second application of the hemoglobin samples directly on top of the first. Proceed *immediately* to electrophoresis.

ELECTROPHORESIS

1. Remove the cover from the electrophoresis chamber.
2. Place the plate, *cellulose acetate side down*, on the wicks, with the samples closer to the *cathodic* (−) buffer reservoir. Set a clean glass slide on the plate (as a weight) to ensure solid contact with the wicks. Up to three plates can be electrophoresed simultaneously in the same chamber.
3. Cover the chamber and then plug in and turn on the power supply.[†] Electrophorese at 350 volts for 25 min.

STAINING

All staining materials should be kept on Benchkote or paper towels.

1. When electrophoresis is completed and the power supply switched off, unplug the power supply and remove the cover from the electrophoresis chamber.

---

*CAUTION: *The hemoglobin samples contain a small quantity of cyanide.*
[†]CAUTION: *While the power supply is turned on, do not touch the electrophoresis chamber or any of the electrical connections.*

2. Remove the cellulose acetate plate from the chamber and place it in a staining rack.
3. Place the rack in a staining dish containing ponceau S for 5 min.
4. Transfer the rack through three destaining baths of 5% acetic acid, 2 min in each bath. The rack should be agitated occasionally in each bath.
5. Dry the plate, *cellulose acetate side up*, in a 56° C oven for 10 min or until dry. Place the dried plate in a protective, clear plastic envelope since cellulose acetate chips rather easily.

REFERENCES

Bunn, H. F., Forget, B. G., and Ranney, H. M. 1977. *Human Hemoglobins*. W. B. Saunders, Philadelphia.

Gaál, Ö., Medgyesi, G. A., and Vereczkey, L. 1980. *Electrophoresis in the Separation of Biological Macromolecules*, pp. 15–18, 35–36, 41–45, 255, 300–307. John Wiley, New York.

Whitaker, J. R., 1967. *Paper Chromatography and Electrophoresis*, Vol. I, pp. 1–49, 102–194. Academic Press, New York.

EXERCISES AND QUESTIONS

1. Identify the band for Hb A and the band for Hb A_2 in sample 1 (Hb A is present in greater quantity than Hb A_2). Also identify the point of application and the band for carbonic anhydrase.

2. Identify the band for Hb S in sample 2.

3. Identify the bands for Hb F and Hb C in sample 3.

4. For each of the five hemoglobins, calculate the electrophoretic mobility (μ). Assume that the distance between the electrodes is the distance between the points where the plate contacted the wicks and that the voltage across this distance is 350 volts. To obtain the distance traveled by each hemoglobin, measure the distance from the point of application to the center of each band. *Remember to include the sign of* μ. Enter the values on Data Sheet 6.1 and show your calculations.

5. From the known amino acid differences in the β chains of hemoglobins A, S, and C (Table 6.1), give an explanation for their relative electrophoretic mobilities.

_____

_____

_____

_____

6. Are the relative electrophoretic mobilities of hemoglobins A, F, S, and C

consistent with their pI's (Table 6.1)? _____ Explain.

_____

_____

_____

_____

7. Give a possible explanation for the fact that Hb A_2 and Hb C have different pI's (Table 6.1) yet have nearly identical electrophoretic mobilities.

_____

_____

_____

8. From the known amino acid differences in hemoglobins A, S, and C, and from your knowledge of the different levels of protein structure, speculate why Hb C results in a less severe condition than does Hb S.

_____

_____

_____

_____

_____

_____

Data Sheet 6.1
Electrophoretic mobilities of hemoglobins A, A$_2$, F, S, and C.

Distance between electrodes = ——————— cm

Time = ——————— sec

Hemoglobin	Distance traveled (cm)	μ (cm^2 sec^{-1} volt^{-1})
A		
A$_2$		
F		
S		
C		

ELECTROPHORESIS OF SERUM PROTEINS

INTRODUCTION

The serum of vertebrates is of interest to molecular biologists because of the many proteins it contains. Serum is the fluid that remains after all cells, platelets, and fibrinogen have been removed from blood. The proteins in serum, which number over 100, have a wide variety of functions including mediation of the immune response, transport of various substances, conservation of iron, and osmotic regulation of blood volume. The best method for separating the serum proteins is electrophoresis; the first such separation was carried out in a liquid medium by A. Tiselius in the 1930s. Today, such separations are done routinely with zone electrophoresis, where there is a solid support medium, like cellulose acetate or a gel. With cellulose acetate electrophoresis, the serum proteins separate into at least four fractions. The objective of this project is to separate the major protein fractions in the sera of four mammals (calf, goat, guinea pig, horse) with cellulose acetate electrophoresis. The percentage of material in each fraction will then be determined by densitometry.

SERUM PROTEINS

Serum proteins separate into four fractions during electrophoresis. From anode to cathode on the electropherogram, the bands are designated *albumin*, *α-globulin*, *β-globulin*, and *γ-globulin*. The amount of material in each band can be measured with a photometric device known as a *densitometer*. The densitometer yields a tracing with a series of peaks, one for each band on the electropherogram. The area under each peak on the *densitometer tracing* is proportional to the amount of material in the corresponding band on the electropherogram. Figure 7.1 shows a densitometer tracing of an electropherogram of rabbit serum. The method for quantifying the material in each band is discussed in the section on Densitometry.

Except for the albumin fraction, which is rather homogeneous, each of the other fractions contains a variety of proteins. Table 7.1 lists most of the common serum proteins in mammals, including humans, and summarizes their functions. Analysis of serum protein levels is of great value clinically, since detection of aberrant patterns can facilitate the diagnosis and management of many diseases in humans and domestic animals.

Albumin.   Depending on the species, albumin constitutes anywhere from 35% to 67% of the total serum proteins, making it the most prominent protein in the liquid portion of the blood. Thus, the albumin fraction is represented by the largest peak in the densitometer tracing (Fig. 7.1). Albumin is one of the smallest serum proteins,

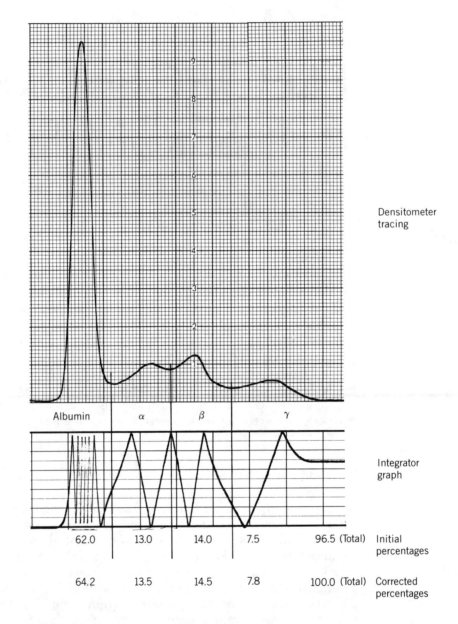

Densitometer
tracing

Integrator
graph

	62.0	13.0	14.0	7.5	96.5 (Total)	Initial percentages
	64.2	13.5	14.5	7.8	100.0 (Total)	Corrected percentages

FIGURE 7.1
Densitometer tracing of an electropherogram of rabbit serum proteins. The four peaks correspond to the four fractions: albumin, α-globulin, β-globulin, and γ-globulin. Below the densitometer tracing is the integrator graph, from which the percentages of material in each fraction can be calculated. See text for explanation.

approximately 580 amino acid residues (65,000–70,000 daltons) in a single polypeptide chain. It is one of the few serum proteins that has no carbohydrate component. Its small size, together with its strong negative charge, gives albumin one of the highest electrophoretic mobilities of the common serum proteins. For a discussion of electrophoretic mobility, see Project 6, page 68. Albumin functions in the transport of diverse substances from the blood to various organs and in the osmotic regulation of blood volume.

α-Globulins. In many animal species, the serum proteins in the α-globulin fraction migrate in two groups, a fast-migrating fraction and a slow-migrating fraction,

TABLE 7.1
Some common serum proteins and their functions.

Serum Protein	Function
Albumin	Osmotic regulation of blood volume; transport of fatty acids and steroid hormones
α-*Globulins*	
α-Lipoproteins	Transport of lipids
α_1-Acid glycoprotein	Function unknown; high carbohydrate content
α_1-Antitrypsin	Inhibits trypsin and related proteases
α_2-Macroglobulin	Inhibits all classes of proteases
Haptoglobins	Conservation of iron; metabolism of hemoglobin
β-*Globulins*	
β-Lipoproteins	Transport of lipids
Transferrin	Transport of iron
Hemopexin	Disposal of heme
Complement proteins	Function with antibodies in removal of antigens
γ-*Globulins* —immunoglobulins	Mediate the immune response

designated α_1-*globulins* and α_2-*globulins*, respectively. The number of α-globulin bands depends both on the support medium and, of course, on the species. With the electrophoresis system that you will be using, rabbit serum has only one α-globulin peak, as shown in Figure 7.1. The various proteins in the α-globulin fraction (or fractions) are related only by their electrophoretic mobility. Table 7.1 shows that the α-globulin fraction contains a diverse group of proteins with a wide variety of functions (lipid transport, enzyme inhibition, iron conservation). The prefix α_1 or α_2 simply denotes whether a protein migrates with the α_1- or α_2-globulin fraction.

β-Globulins. In many animal species, the proteins in the β-globulin fraction also migrate in two groups, designated β_1-*globulins* (fast) and β_2-*globulins* (slow). Again, in the rabbit, there is only one peak (Fig. 7.1). Like α-globulins, β-globulins also contain proteins with a wide array of functions (Table 7.1). These include lipid transport, hemoglobin metabolism, and defense against microorganisms. This last function is effected by the *complement proteins*, which constitute about 10% of the serum proteins in all vertebrates. These proteins are aptly named, for they "complement" the function of antibodies in the removal of *antigens*, i.e., the bacteria and foreign particles that stimulate the production and release of antibodies.

γ-Globulins. The proteins in the γ-globulin fraction may separate as one or two bands, depending on the species and the electrophoresis system. The γ-globulin fraction is of special interest because it contains the antibodies, or *immunoglobulins*. Since all of the immunoglobulins have related amino acid sequences, this fraction often appears as a broad band on the electropherogram and has a correspondingly wide peak on the densitometer tracing (Fig. 7.1).

CELLULOSE ACETATE ELECTROPHORESIS OF SERUM PROTEINS

In this project, cellulose acetate electrophoresis will be used to separate the major fractions in calf, goat, guinea pig, and horse sera. The basis for the separation is the different electrophoretic mobilities, i.e., rates of migration, of the various serum proteins. Electrophoretic mobility, along with the theory of electrophoresis, is discussed in Project 6, pp. 66–68, and should be reviewed. Analysis of the electropherogram, i.e., the stained protein bands on the cellulose acetate plate, will facilitate a comparison of the serum fractions in the four species.

The components of the electrophoresis system are depicted in Figure 6.4 (p. 66) of the preceding project. Buffer at pH 8.6–9.0 is added to the outer anodic and cathodic buffer reservoirs, and a buffer-soaked wick is draped over the inner wall of each buffer reservoir. After the cellulose acetate plate has been saturated with buffer, the serum samples are applied (at the center of the plate) using the applicator depicted in Figure 6.5 (p. 68). The serum samples are electrophoresed at 180 volts for 15 min. *The high voltage used for electrophoresis demands that extreme caution be exercised; never touch the electrophoresis chamber or any of the electrical connections while the power is turned on.*

After electrophoresis, the cellulose acetate plate is placed in a solution of ponceau S, which stains the proteins. A dilute solution of acetic acid is used to remove background staining and then the cellulose acetate is *cleared*, i.e., made transparent, with polyethylene glycol. Clearing is necessary so that the electropherogram can be scanned with a densitometer.

From visual observation of the electropherogram, it is not easy to distinguish the five bands that are actually present in each sample. The albumin fraction appears as a large, discrete band at the anodic end of the electropherogram. The other bands are not as distinct and may appear as a continuous smear. Further resolution of the stained bands can be achieved with a scanning densitometer.

DENSITOMETRY

A densitometer is an instrument that quantifies the material in the bands of the electropherogram. Most densitometers are of the transmittance type, which measures sample absorbance with a beam of light that passes through the support medium. The support medium must, therefore, be transparent so that all of the incident light can pass through. The stained bands absorb light in proportion to the concentration of material they contain. To maximize absorbance of the stained bands, the color of the incident light must be complementary to that of the stain. In this case, a green filter (525 nm) is used to complement the red staining of the serum protein bands. When the beam passes through a band, some of the light is absorbed by the stained material and the remainder is transmitted. The transmitted light is measured by a photocell, which converts the light energy into an electric current. The electric current is then converted by the circuitry of the instrument into an absorbance reading.

The densitometer scans the electropherogram by moving it past the narrow beam of light at a constant speed. The instrument records the absorbances in the form of a tracing on graph paper (Fig. 7.1). Each band present on the electropherogram appears as a *peak* on the tracing, with the area under each peak representing the total absorbance of the corresponding band. Since absorbance is directly proportional to concentration (a principle known as Beer's law), the relative area under each peak also represents the relative amount of material present in the corresponding band. In the more sophisticated densitometers, the percentages of material in the bands are printed out directly. In simpler instruments, the operator must determine these values from the *zigzag, time-base integrator graph*, which is generated simultaneously and directly beneath the densitometer tracing (Fig. 7.1).

Identification of the Fractions from the Densitometer Tracing.   In this project, every densitometer tracing will have five clearly discernible peaks, corresponding to the five bands on the electropherogram. Data Sheet 7.2 lists, for each species, the separated fractions in sequence, from anode to cathode. In calf, goat, and guinea pig, there are two α-globulin peaks and a single β-globulin peak. In the horse, there is a single α-globulin peak and two β-globulin peaks. In all four species, there is a single albumin peak and a single γ-globulin peak.

Determining the Relative Area under Each Peak. The first step in determining the percentage of material in each fraction is to identify the boundaries between adjacent peaks on the densitometer tracing. A boundary is the lowest point, or "valley," between adjacent peaks. After the boundaries have been identified on the densitometer tracing, they must be marked off on the integrator graph below. This has been done in Figure 7.1, where the boundaries between peaks have been marked with vertical lines that extend from the low points on the tracing all the way through the integrator graph.

Next, the bold, horizontal lines traversed by the pen on the integrator graph are counted. Each bold line traversed represents 1% of the total material. Since there are five bold lines from top to bottom on the integrator graph paper, each complete up-and-down movement of the pen represents 10% of the total. In Figure 7.1, for example, the relative area under the albumin peak was calculated to be 62.0%, since six complete up-and-down lines (60%) plus 2.0 additional bold, horizontal lines (2.0%) are traversed. After the relative areas are obtained for all the peaks, these initial percentage values should be totaled. If they total to less than 100%, the percentages must be corrected. For example, in this case, they total to 96.5%, so each percentage value must be divided by 0.965 to obtain corrected values that do total to 100% (62.0%/0.995 = 64.2%, etc.).

PROCEDURES[1]

For all procedures, you will work in teams of four. Each team will electrophorese four serum samples: calf, goat, guinea pig, horse. Each team will have its own micropipetter, sample well plate, applicator, and cellulose acetate plate.

PREPARATION OF THE ELECTROPHORESIS CHAMBER

1. After making certain that the power supply is turned off and unplugged, add 100 ml of buffer* to each outer reservoir (anodic and cathodic) of the electrophoresis chamber.
2. Wearing disposable gloves, moisten two wicks with buffer. Then, drape a wet wick over the inner wall of each buffer reservoir, making certain that each wick is immersed in the buffer.
3. Cover the chamber to prevent evaporation.

APPLICATION OF THE SERUM SAMPLES

Care should be taken when handling the serum samples since there is always the slight possibility that they are contaminated with infectious agents. Accordingly, disposable gloves should be worn, all work surfaces covered with Benchkote or paper towels, spills wiped up, the area disinfected, and all waste materials disposed of properly.

1. Familiarize yourself with the following equipment items: the Bufferizer, a set of two containers designed to immerse the cellulose acetate plate into buffer very slowly; the micropipetter and disposable pipet tips; the sample well plate, the reservoir for the serum samples; the applicator; and the aligning base, the

---

[1]The electrophoresis procedure has been adapted from a protocol developed by Helena Laboratories and is used here with permission.

*CAUTION: *The buffer is toxic; wash thoroughly after handling.*

plastic holder that positions the cellulose acetate plate during sample application.

2. With a permanent marking pen, make a small identification mark at a corner of the cellulose acetate plate, on the Mylar (glossy) side.

3. Place the plate in a rack and then place the rack in the lower chamber of the Bufferizer. Add buffer to the upper chamber and proceed to step 4 while the plates are soaking. The plates should soak about 20 min (after they are completely immersed in buffer) before the samples are applied.

4. Using the micropipetter, place 4.0 µl of each serum in a separate well of the sample well plate. Run the separations in duplicate, adding the calf serum to wells 1 and 2, the goat serum to wells 3 and 4, the guinea pig serum to wells 5 and 6, and the horse serum to wells 7 and 8. The easiest way to fill each well is to expel the drop at the pipet tip and then touch the drop to the well. Be sure to use a separate pipet tip for each sample. To prevent evaporation of the samples, cover the sample wells with a clean glass slide.

5. After the plates have been soaking 20 min, prime the applicator tips as follows. Place the applicator in the brackets of the sample well plate and depress the tips three or four times in the serum samples. Remove the applicator from the brackets and touch the tips to a blotter. Replace the applicator in the brackets of the sample well plate and then proceed quickly through the next three steps.

6. Remove the cellulose acetate plate from the soaking buffer and place it on a blotter, *cellulose acetate side up*. With another blotter, blot the cellulose acetate once firmly.

7. Place a drop of water on the aligning base and then place the plate, *cellulose acetate side up*, in the aligning base so that the bottom edge of the plate is on the line labeled CENTER APPLICATION. In this position, the samples will be applied at the center of the plate. One person should apply the serum samples for each team.

8. Apply the serum samples as follows, keeping in mind that once the applicator tips are loaded, the samples must be applied within 15 sec. Load the applicator tips by depressing them into the serum samples three or four times. Immediately place the applicator in the brackets of the aligning base, which contains the cellulose acetate plate. Depress the tips, and hold for 5 sec. Proceed *immediately* to electrophoresis.

ELECTROPHORESIS

1. Remove the cover from the electrophoresis chamber.

2. Place the plate, *cellulose acetate side down*, on the wicks, with the identification mark closer to the anode. Set a clean glass slide on the plate (as a weight) to ensure solid contact with the wicks. Up to three plates can be electrophoresed simultaneously in the same chamber.

3. Cover the chamber, wait 30 sec, and then plug in and turn on the power supply.* Electrophorese at 180 volts for 15 min.

STAINING AND CLEARING

All materials should be kept on Benchkote or paper towels and disposable gloves worn.

*CAUTION: *While the power supply is turned on, do not touch the electrophoresis chamber or any of the electrical connections.*

1. When electrophoresis is completed and the power supply switched off, unplug the power supply and remove the cover from the electrophoresis chamber.
2. Remove the cellulose acetate plate from the chamber and place it in a staining rack.
3. Place the rack in a staining dish containing ponceau S for 6 min.
4. Transfer the rack through three destaining baths of 5% acetic acid, 2 min in each bath. The rack should be agitated occasionally in each bath.
5. Transfer the rack through two dehydrating baths of absolute methanol, 2 min in each bath.
6. Drain the rack on paper towels for 10 sec and then place the rack into the clearing solution (30 parts glacial acetic acid, 70 parts absolute methanol, 4 parts polyethylene glycol) for 10 min.
7. Drain the rack on paper towels for 10 sec. Then place each plate, *cellulose acetate side up*, on a blotter and dry in a 56° C oven for 15 min or until dry. For storage, place the dried plate in a protective, clear plastic envelope.

DENSITOMETRY

Following the instructions for the densitometer available in your lab, obtain a densitometer tracing for the electropherogram. For each species, select the better of the two lanes for scanning (measurement). Do not select a lane with curved or torn bands. Scan the plate using a 525-nm filter.

REFERENCES

Gaál, Ö., Medgyesi, G. A., and Vereczkey, L. 1980. *Electrophoresis in the Separation of Biological Macromolecules*, pp. 178–181, 307–335. John Wiley, New York.

Hood, L. E., Weissman, I. L., Wood, W. B., and Wilson, J. H. 1984. *Immunology*, 2nd ed., pp. 334–348. Benjamin/Cummings, Menlo Park, CA.

Kaneko, J. J. 1980. Serum proteins and the dysproteinemias. In *Clinical Biochemistry of Domestic Animals*, 3rd ed., Kaneko, J. J., ed., pp. 97–118. Academic Press, New York.

Putnam, F. W. 1975. Alpha, beta, gamma, omega—The roster of the plasma proteins. In *The Plasma Proteins*, 2nd ed., Vol. 1, Putnam, F. W., ed., pp. 58–131. Academic Press, New York.

Smith, E. L., Hill, R. L., Lehman, I. R., Lefkowitz, R. J., Handler, P., and White, A. 1983. *Principles of Biochemistry: Mammalian Biochemistry*, 7th ed., pp. 5–16, 38–66. McGraw–Hill, New York.

EXERCISES AND QUESTIONS

1. Place the electropherogram on a piece of white paper (to facilitate visualization of the bands). From visual examination alone, for which, if any, of the four species are five bands evident on the electropherogram?

_____

2. Examine the densitometer tracing. Does the curve go to the baseline between

   the peaks? _____ Explain.

   _____

   _____

3. For each species, calculate the electrophoretic mobility (μ) of the albumin fraction. The equation for electrophoretic mobility is given in Project 6, page 68. Assume that the distance between the electrodes is the distance between the points where the cellulose acetate plate contacted the wicks and that the voltage across this distance is 180 volts. To obtain the distance traveled by the albumin fraction, measure the distance from the point of application to the center of the albumin band. *Remember to include the sign of* μ. Enter the values on Data Sheet 7.1 and show your calculations.

4. Give a possible explanation for differences in the electrophoretic mobility of the albumin fraction in the different species.

   _____

   _____

   _____

5. Examine the densitometer tracings of all four species. Using Data Sheet 7.2 as a guide, identify each peak. Then, under each peak, write the name of the corresponding fraction.

6. Examine the α-globulin peaks in calf, goat, and guinea pig. In each species, which is larger, α_1 or α_2?

   _____

   _____

   _____

7. Examine the β-globulin peaks in the horse. Which is larger, β_1 or β_2? _____
8. In which of the four species is the γ-globulin peak the widest of the five peaks?

   _____

9. From the densitometer tracings and integrator graphs, determine the percentage of each serum fraction in the four species. Enter the values on Data Sheet 7.2.

10. Compare the percentages of the serum protein fractions in the four species. Describe any trends that are evident.

_____

_____

_____

_____

Data Sheet 7.1
Electrophoretic mobility of the serum albumin fraction in calf, goat, guinea pig, and horse.

Distance between electrodes = _____ cm

Time = _____ sec

Species	Distance traveled (cm)	$\mu\,(cm^2\,sec^{-1}\,volt^{-1})$
Calf		
Goat		
Guinea pig		
Horse		

Data Sheet 7.2

Percentage of each serum fraction in calf, goat, guinea pig, and horse.

Species	Percentage				
Calf	Albumin	α_1	α_2	β	γ
Goat	Albumin	α_1	α_2	β	γ
Guinea pig	Albumin	α_1	α_2	β	γ
Horse	Albumin	α	β_1	β_2	γ

ISOZYME PATTERNS OF LACTATE DEHYDROGENASE

INTRODUCTION

Within the cells and tissues of the same organism, certain enzymes occur in multiple molecular forms. Such multiple forms of an enzyme are termed *isoenzymes* or *isozymes*. One of the best studied families of isozymes is *lactate dehydrogenase* (*LDH*), which exists in five common forms. LDH catalyzes the final reaction of anaerobic glycolysis: the reduction of pyruvate to lactate and the concomitant oxidation of NADH to NAD^+ (*nicotinamide adenine dinucleotide*). Each of the five LDH isozymes catalyzes the same reaction, but each has unique kinetic properties. The pattern of the five LDH isozymes in a particular tissue plays a significant role in its metabolism by determining the rate at which pyruvate is converted to lactate. The objective of this project is to determine the relative amounts of the five LDH isozymes in extracts of several different tissues of the rat. The method used will be electrophoresis with an *agarose gel* as the support medium.

THE ISOZYMES OF LACTATE DEHYDROGENASE

Each of the five major LDH isozymes consists of four polypeptide chains. Biochemical studies have revealed that each polypeptide can be of two types, designated the *M* and *H* chains. These designations stem from the predominance of the M chain in the LDH isozymes of skeletal muscle and the predominance of the H chain in the LDH isozymes of the heart. Since there are four chains in each isozyme molecule and two possible subunits (M or H), there are five possible combinations. When all four chains are of the H type, the resultant tetramer has the formula H_4, also referred to as *LDH-1*. The LDH isozyme with one M chain and three H chains has the formula MH_3 and is designated *LDH-2*. Similarly, the LDH isozymes with the other combinations of the M and H chains have the formulas M_2H_2 (*LDH-3*), M_3H (*LDH-4*), and M_4 (*LDH-5*).

Each LDH isozyme has a molecular weight of about 140,000 daltons, and each polypeptide chain has a molecular weight of about 35,000 daltons. Despite their similar molecular weights, the M and H chains have very different amino acid compositions and confer different kinetic properties on the LDH molecule. The rate of conversion of pyruvate to lactate increases with the number of M chains. Thus, M_4 (LDH-5) has the highest rate for converting pyruvate to lactate, while H_4 (LDH-1) has the lowest rate and is even inhibited by excess pyruvate. The other LDH isozymes have rates that are proportional to the number of M chains.

The relative amounts of the five LDH isozymes in a tissue correlate well with the metabolic requirements of the tissue. In skeletal muscle, where oxygen deprivation is a common occurrence, LDH-5 is the most abundant isozyme. Since LDH-5 can rapidly convert pyruvate to lactate, the predominance of this isozyme allows glycolysis and ATP synthesis to continue in the absence of oxygen. In cardiac muscle, which is a more aerobic tissue, LDH-1 and LDH-2 predominate. Since these isozymes convert pyruvate to lactate very slowly, the pyruvate is preferentially oxidized *aerobically* to CO_2 and H_2O.

Different patterns of the five LDH isozymes are also observed at different times during embryonic development. In the rat and mouse, for example, most early embryonic tissues, including heart muscle, have LDH-5 and LDH-4 as the predominant isozymes. As development proceeds there is a gradual shift in many tissues toward the isozymes with more H chains.

The relative amounts of the five LDH isozymes at a specific time during development or in a particular tissue are determined primarily by the proportions of the M and H subunits in each cell. It has been shown that if the single M and H chains are mixed in specific proportions *in vitro*, they aggregate to form the different LDH isozymes in proportions predicted by the laws of probability. Like all proteins, the synthesis of the M and H chains is directed by genes. In cells where there is a preponderance of one of the two chains (M or H), we have an example of differential gene activity, with the genes that encode the M and H chains being transcribed at different rates. There is also evidence that during cell differentiation, different LDH isozymes are selectively degraded in certain tissues.

Assays of LDH isozymes are important not only in studies of metabolic regulation and cell differentiation but in clinical evaluations, as well. For example, normally LDH-2 is the predominant LDH isozyme in human serum, with LDH-1 being the second most abundant form. Following a heart attack, however, the damaged heart tissue releases excess LDH-1, so that LDH-1 becomes the predominant LDH isozyme in the blood. Such enzyme assays are used not only for evaluating cardiovascular disease but as an aid in the diagnosis of blood and liver disorders.

PREPARATION OF THE TISSUE EXTRACTS

You will be separating the LDH isozymes from the heart, liver, and gastrocnemius (calf) muscle of the rat. The cell fractions can be prepared in the laboratory by thoroughly mincing each tissue and then homogenizing with cold distilled water in a motor-driven, *Potter–Elvehjem tissue grinder* (Fig. 8.1), which is a glass tube with a fitted Teflon (Du Pont) or glass pestle. As the rotating pestle moves toward the bottom of the tube, the tissue is forced upward into the narrow space between the pestle and the wall of the tube. The shearing forces that are generated in this process disrupt the plasma membrane and release the cell contents. Care must be taken that the tissue does not get too warm during the homogenization since heating could destroy enzymatic activity, particularly of LDH-5. Accordingly, during homogenization, the tube should be immersed periodically in an ice-water bath. The resultant homogenate

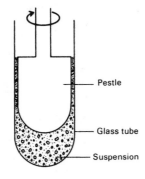

FIGURE 8.1
Potter–Elvehjem tissue grinder for homogenizing animal tissue. The shaft of the pestle is connected to the shaft of a motor.

ISOZYME PATTERNS OF LACTATE DEHYDROGENASE

is centrifuged to sediment tissue debris. The supernatant (suspended particles and liquid above the pellet) contains the LDH isozymes along with many other cellular molecules. Before it is applied to the support medium for electrophoresis, the supernatant must be diluted. The diluted supernatant will hereafter be referred to as the *tissue extract*.

AGAROSE GEL ELECTROPHORESIS

The principles underlying the electrophoretic separation of the LDH isozymes are the same as discussed in Project 6, page 66, for hemoglobins. The differences in the two separations are strictly procedural. The electrophoresis chamber and power supply are the same as in Project 6, but the setup within the electrophoresis chamber is different because of the different support medium. For the separation of LDH isozymes in this project, the support medium is a thin layer of agarose gel affixed to a sheet of Mylar (Du Pont). The agarose plate yields an electropherogram that is easily handled and has a transparent background for densitometry (discussed in Project 7, pp. 80–81). Agarose is a polysaccharide that forms a porous gel, the diameter of the pores being a function of the agarose concentration. It is purer and much less acidic than ordinary agar, from which it is derived.

As with other types of electrophoresis, each sample must be applied to the support medium as a narrow streak. To apply the samples to the agarose gel, we use a cellophane template with narrow slits. With the template positioned on the agarose, several microliters of the diluted supernatant are placed in each slit (Fig. 8.2). After the samples have had time to diffuse into the agarose, the plate is ready for electrophoresis. The ends of the agarose gel plate will be immersed *directly* in the buffer (pH 8.1-8.3). Accordingly, the buffer is added to the *inner* buffer reservoirs of the electrophoresis chamber, and the agarose gel plate is flexed so as to fit in these wells. Thus, wicks are not required (as they were for the cellulose acetate plates in the separations of hemoglobins and serum proteins). The plate should be oriented with the *agarose side down* and the samples closer to the *cathodic* buffer reservoir. The chamber is covered and the samples electrophoresed at 100 volts for 15 minutes. *The high voltage used for electrophoresis demands that extreme caution be exercised; never touch the electrophoresis chamber or any of the electrical connections while the power is turned on.* Each of the five LDH isozymes has a different electrophoretic mobility, with the final sequence from anode (+) to cathode (−) being LDH-1, LDH-2, LDH-3, LDH-4, LDH-5.

VISUALIZATION AND QUANTIFICATION OF THE LDH BANDS

The tissue extract applied to the agarose gel contains not only the five LDH isozymes but many other proteins and cellular molecules, all colorless. To identify the LDH bands, we selectively stain them after electrophoresis. The agarose plate is incubated with a reaction mixture containing lithium lactate, NAD$^+$, nitro blue tetrazolium, and phenazine methosulfate. The lithium lactate serves as a substrate for the enzyme.

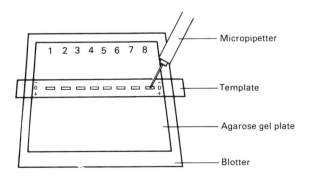

FIGURE 8.2

Application of samples to the agarose gel plate. See text for explanation.

Wherever there is LDH activity in the agarose gel, the lactate in the reagent is converted to pyruvate, and the NAD^+ is reduced to NADH, as summarized by

$$\text{lactate} + NAD^+ \xrightarrow{\text{LDH}} \text{pyruvate} + NADH + H^+$$

The hydrogens from the reduced NAD are transferred by phenazine methosulfate to nitro blue tetrazolium, reducing it to an insoluble, purple formazan dye, as follows:

$$NADH + H^+ + \text{nitro blue tetrazolium} \xrightarrow[\text{methosulfate}]{\text{phenazine}} NAD^+ + \underset{\text{(purple)}}{\text{formazan dye}}$$

Thus, each LDH band will be stained purple. To remove background staining, the agarose plate is soaked in a dilute solution of acetic acid. The agarose gel is then blotted with an acetic acid–moistened blotter and dried in an oven.

The foregoing staining reaction is stoichiometric; i.e., the intensity of the purple color is proportional to the amount of LDH present. Any striking differences in staining intensity among the five bands can be readily detected by visual examination.

DENSITOMETRY

For accurate quantification of the LDH present in each band of the electropherogram, a densitometer is required. The principles of densitometry are discussed in Project 7, pages 80–81, and should be reviewed if a densitometer will be used here. In this project, the electropherogram is scanned at 595 nm for the purple-stained LDH bands. As the transmitted light strikes a stained region (band) in the otherwise transparent agarose gel, the densitometer measures photoelectrically and records graphically how much light is absorbed by the band. In the final graph (Fig. 8.3), there is a series of steep curves, or peaks, one for each band. The area under each of the five peaks is proportional to the amount of LDH in the corresponding band. Measurement of the areas under the five peaks in Figure 8.3 indicates that the relative amounts of LDH-1, LDH-2, LDH-3, LDH-4, and LDH-5 are 5.2%, 5.7%, 4.7%, 1.6%, and 82.8%, respectively, in a commercially prepared extract of rat liver.

PROCEDURES[1]

PREPARATION OF THE TISSUE EXTRACTS

For the entire project, you should work in teams of four. Each team will be responsible for obtaining one of the tissue extracts for the entire class. Every team will then be able to electrophorese extracts from four different tissues: heart, liver, gastrocnemius muscle, and one other tissue. For all tissue homogenization procedures, the solutions and containers must be ice-cold to maintain the integrity of the lactate dehydrogenase.

1. Sacrifice an adult rat with a sharp blow to the head or by decapitation.
2. Wearing disposable gloves, dissect out the organ that has been assigned to your team: heart, liver, gastrocnemius muscle, spleen, kidney, or lung. For all procedures, each tissue is to be kept separate from the others.

---

[1]The electrophoresis procedure has been adapted from a protocol developed by Helena Laboratories and is used here with permission.

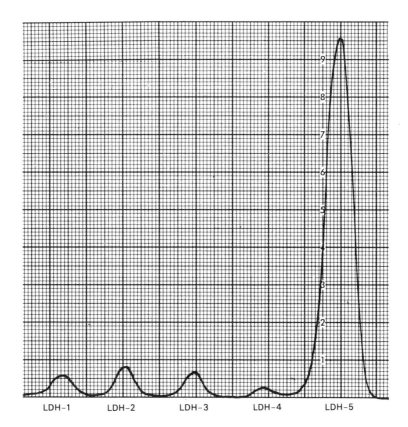

FIGURE 8.3
Densitometer tracing of an electropherogram of the LDH isozymes in rat liver.

LDH–1   LDH–2   LDH–3   LDH–4   LDH–5

3. Rinse the tissue with ice-cold distilled water to remove as much blood as possible. Drain and weigh the tissue, recording the weight. Use no more than 2 g of tissue.

4. Place the tissue on a wooden chopping block and mince the tissue finely with a single-edge razor blade.

5. Place the minced tissue in the tube of a motor-driven, Potter–Elvehjem tissue grinder and add 1.0 ml of ice-cold distilled water for each gram of tissue.

6. Start the motor and then, with the rotating pestle in the tube (Fig. 8.1), move the tube up and down. To prevent the tissue from getting too warm during the homogenization, periodically place the tube in an ice-water bath. Continue the process until the tissue is homogenized thoroughly, or as much as seems possible.

7. Transfer the homogenate to a centrifuge tube and centrifuge at 1300 g for 10 min at 0°–4° C. Make sure that the centrifuge tubes are balanced; the tubes opposite each other should have the same total volume.

8. Transfer 1.0 ml of the supernatant* to an Erlenmyer flask and dilute with 49 ml of ice-cold distilled water. Mix well and then dispense the diluted supernatant (tissue extract) in a small vial to each team.

PREPARATION OF THE ELECTROPHORESIS CHAMBER

1. After making certain that the power supply is turned off and unplugged, add 35 ml of buffer† to each inner reservoir (anodic and cathodic) of the electrophoresis chamber.

*CAUTION: *Do not pipet by mouth.*
†CAUTION: *The buffer is toxic; wash thoroughly after handling.*

2. Cover the chamber to prevent evaporation.

APPLICATION OF THE TISSUE EXTRACTS

Each team will have its own micropipetter and agarose gel plate.

1. Place a gel plate,* *agarose side up*, on a blotter D. The numbers 1–8 (for the eight samples) should be readable across the top of the plate. With a permanent marking pen, make a small identification mark at the corner of the plate, on the Mylar (glossy) side.

2. With a blotter A, gently blot the agarose where the samples will be applied, i.e., between the $\bar{\underline{0}}_+$ demarcations. Then discard blotter A.

3. Place a cellophane template on the agarose so that the template slits, which will serve as sample wells, are aligned with the zero (0) demarcations at the sides of the plate (see Fig. 8.2). Run your fingertip lightly across the template to press out any air bubbles.

4. Using a micropipetter, place 3.0 μl of each tissue extract in a separate template slit (see Fig. 8.2). Run the separations in duplicate, adding the tissue extract from heart to slits 1 and 2, the tissue extract from liver to slits 3 and 4, the tissue extract from muscle to slits 5 and 6, and the tissue extract from the fourth tissue to slits 7 and 8. The easiest way to fill each slit is to expel the drop at the pipet tip and then touch the drop to the slit. Use a fresh pipet tip for each tissue extract and be sure to record which sample is in which numbered slit. Work carefully and quickly.

5. After the last sample has been dispensed, wait 3 min; then gently blot the template with a fresh blotter A.

6. Remove and discard blotter A, wait 30 sec, and then remove and discard the template and blotter D. Proceed *immediately* to electrophoresis.

ELECTROPHORESIS

1. Wearing disposable gloves, remove the cover from the electrophoresis chamber.

2. With the *agarose side down* and the samples closer to the *cathodic* (−) buffer reservoir, place the ends of the gel plate into the inner buffer reservoirs, curling the plate so that it will fit. Two plates can be electrophoresed simultaneously in the same chamber.

3. Cover the chamber, making certain that the cover does not touch the gel plate(s), and then plug in and turn on the power supply.[†] Electrophorese at 100 volts for 15 min.

VISUALIZATION OF THE LDH BANDS

1. Immediately after electrophoresis is begun, prepare an incubation chamber (plastic case) as follows. Place a paper towel moistened with water in the incubation chamber and then place the closed chamber in a 45°C incubator until needed in step 7.

2. When electrophoresis is completed and the power supply switched off, unplug the power supply and then remove the cover from the electrophoresis chamber.

---

*__CAUTION:__ *The agarose gel plate contains sodium azide, which is poisonous. When handling the plate, keep hands away from mouth. After handling the plate, wash your hands.*

[†]__CAUTION:__ *While the power supply is turned on, do not touch the electrophoresis chamber or any of the electrical connections.*

ISOZYME PATTERNS OF LACTATE DEHYDROGENASE

3. Place a clean glass development slide (7.6 cm × 10 cm) on a paper towel.

4. Remove the gel plate from the electrophoresis chamber and place it, *agarose side up*, on the glass development slide. The anode edge of the plate should be closer to you.

5. Pour the entire vial of reconstituted LDH isozyme reagent along the cathode edge of the agarose.

6. Use a 5-ml pipet to spread the reagent across the agarose as follows. Lay the pipet lengthwise along the cathode edge of the plate. Spread the reagent by gently and slowly pulling the pipet toward you, being careful not to roll the reagent off the plate. Wait 15 sec, then reverse the direction of the spreading. Gently and slowly, move the pipet away from you, again being careful not to roll the reagent off the plate. Wait 15 sec and then spread the reagent toward you, this time allowing the reagent to roll off the plate.

7. Place the gel plate, *agarose side up*, in the preheated incubation chamber.

8. Place the closed incubation chamber in a 45° C incubator for 25 min.

9. After the incubation period, transfer the gel plate, *agarose side up*, to a large finger bowl containing 10% acetic acid, 1–2 cm deep, and gently swirl the solution for 2 min.

10. Transfer the gel plate, *agarose side up*, to a clean glass development slide. Then, stack the following on the plate: one blotter B moistened with 10% acetic acid (on the agarose), three folded paper towels (on blotter B), and, to provide pressure, a development weight (on the paper towels).

11. After 5 min, remove the weight, discard the blotter and paper towels, and then dry the gel plate by placing it, *agarose side up*, in a 60° C oven for 5 min or until dry. The electropherogram can now be evaluated visually or with a densitometer (at 595 nm), if one is available. For visual examination and storage, keep the dried plate in a clear plastic envelope.

REFERENCES

Everse, J. and Kaplan, N. O. 1973. Lactate dehydrogenases: structure and function. In *Advances in Enzymology*, Vol. 37, Meister, A., ed., pp. 61–133. John Wiley, New York.

Gaál, Ö., Medgyesi, G. A., and Vereczkey, L. 1980. *Electrophoresis in the Separation of Biological Macromolecules*, pp. 53–65, 178–186. John Wiley, New York.

Markert, C. L. and Ursprung, H. 1971. *Developmental Genetics*, pp. 36–49. Prentice-Hall, Englewood Cliffs, N.J.

Moss, D. W. 1982 *Isoenzymes*. Chapman and Hall, London.

EXERCISES AND QUESTIONS

1. Examine the electropherogram and locate the bands containing LDH-1, LDH-2, LDH-3, LDH-4, and LDH-5.

2. Which LDH band(s) migrated toward the cathode? _____ What are the possible causes for this movement?

   _____

   _____

   _____

3. For the heart, give the sequence of the five isozymes, from highest concentration to lowest concentration.

   _____   _____   _____   _____   _____
   Highest   Lowest

4. For the liver, give the sequence of the five isozymes, from highest concentration to lowest concentration.

   _____   _____   _____   _____   _____
   Highest   Lowest

5. For skeletal muscle, give the sequence of the five isozymes, from highest concentration to lowest concentration.

   _____   _____   _____   _____   _____
   Highest   Lowest

6. Name the fourth tissue and give the sequence of the five isozymes, from highest

   concentration to lowest concentration. _____

   _____   _____   _____   _____   _____
   Highest   Lowest

7. How do the LDH patterns in the heart and skeletal muscle differ?

   _____

   _____

   _____

   _____

   Which of these two patterns (heart or skeletal muscle) does the LDH pattern from

   liver more closely resemble? _____

   Which of these two patterns (heart or skeletal muscle) does the LDH pattern from

   the fourth tissue more closely resemble? _____

8. In which tissue(s) is there primarily only one of the two polypeptide chains (M or

H) present? _____
Suggest events that might be occurring at the molecular level (transcriptional, posttranscriptional) to explain such LDH profiles.

_____

_____

_____

_____

_____

9. If a densitometer is available, determine the percentage of each LDH isozyme in each tissue. Enter the values on Data Sheet 8.1.

NAME _____ SECTION _____ DATE _____

LAB PARTNERS _____

Data Sheet 8.1

Percentage of each LDH isozyme in heart, liver, skeletal muscle, and the fourth tissue (identify in table).

Tissue	LDH-1	LDH-2	LDH-3	LDH-4	LDH-5
Heart					
Liver					
Muscle					

MEMBRANE PERMEABILITY

INTRODUCTION

Every cell is separated from its external environment by a *plasma membrane*. This thin (8.5 nm), primarily lipid–protein structure not only surrounds and contains the living protoplasm but regulates what substances enter and leave the cell. Investigations of the plasma membrane date back to the latter part of the nineteenth century when physiologists discovered its *selective permeability*. They observed that different substances penetrate the cell at different rates and that some substances do not penetrate at all. In 1899, E. Overton demonstrated that the greater the lipid solubility of a compound, the greater is its rate of penetration. This correlation, sometimes referred to as *Overton's rule*, provided one of the earliest indications that lipids are a major component of the plasma membrane. The objective of this project is to test alcohols with different lipid solubilities for their relative abilities to penetrate *Elodea* leaf cells.

MEMBRANE STRUCTURE

Today, the most widely accepted model for membrane structure is the *fluid mosaic model* (Fig. 9.1), proposed by S. J. Singer and G. L. Nicolson in 1972. Here, as in previous hypotheses, there is a *lipid bilayer*, with the hydrophobic, nonpolar "tails" of the phospholipid molecules pointing inward and the hydrophilic, polar "heads" of the phospholipids facing outward. What makes the fluid mosaic model unique is the dynamic relationship between the lipid bilayer and membrane proteins. The proteins are not in static layers, as previous models had proposed, but are completely interspersed with the lipids, producing a mosaic. The model portrays the lipid bilayer as a rather fluid structure in which the various proteins can move laterally. Two classes of proteins are proposed: the *integral proteins*, which penetrate the lipid bilayer partially or completely, and the *peripheral proteins*, which are attached to the surface. Some of the integral proteins are glycoproteins, having carbohydrate chains attached to them. Some of the lipids also have carbohydrate chains attached; these are the glycolipids. The carbohydrate chains always protrude from the cell surface to the exterior and, in certain cell types, confer antigenic specificity to the cell.

OSMOSIS IN PLANT AND ANIMAL CELLS

In many cells, such as *Elodea* leaf cells, the plasma membrane is freely permeable to water. The movement of water into or out of cells occurs by *diffusion*, a process of molecular movement from a region of higher concentration to a region of lower concentration due to the kinetic energy of the molecules. The diffusion of water across a membrane in response to a concentration gradient is termed *osmosis*. When a plant cell is placed in pure water, the water flows inward by osmosis. The cell volume

103

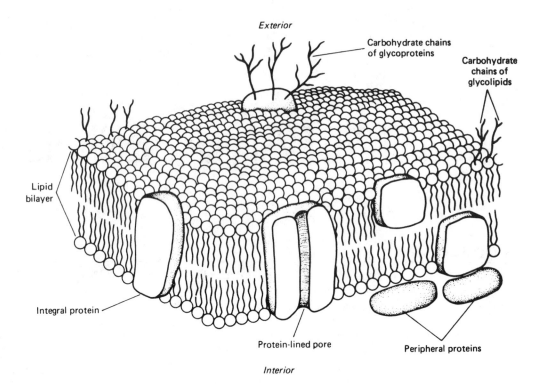

FIGURE 9.1
Fluid mosaic model of membrane structure.

Labels in figure:
Exterior
Carbohydrate chains of glycoproteins
Carbohydrate chains of glycolipids
Lipid bilayer
Integral protein
Protein-lined pore
Peripheral proteins
Interior

increases, causing the plasma membrane to expand against the relatively rigid cell wall. The resultant *osmotic pressure* increases until it prevents any further net flow inward, at which point an equilibrium is reached. When an animal cell such as an erythrocyte is placed in pure water, the consequences are different. As in the plant cell, water enters by osmosis, and the plasma membrane expands. Without a restraining cell wall, however, the membrane of the erythrocyte tears and the cell contents are released, a result known as *hemolysis*.

Both the cytoplasm and the liquid that bathes the cell are aqueous solutions; water is the *solvent*, and the proteins, sugars, and other dissolved material comprise the *solute*. The *solute concentration* is defined as the total number of dissolved particles per unit volume of solution. The terms *hypertonic*, *hypotonic*, and *isotonic* are used to describe relative solute concentrations. If the solute concentration outside the cell is *greater* than that inside, the external solution is described as *hypertonic*. Conversely, if the solute concentration outside the cell is *less* than that inside, the external solution is said to be *hypotonic*. A solution with the *same* solute concentration as the cell is *isotonic*.

Plasmolysis and Deplasmolysis in *Elodea* Leaf Cells. Let us consider the situation of an *Elodea* leaf placed in an aqueous solution in which the solute penetrates the plasma membrane at a negligible rate. One such compound is sucrose. If the leaf is placed in an isotonic sucrose solution, there will be no net flow of water into or out of the cells because the concentration of the water (solvent) is equal inside and outside. If, however, the leaf is placed in a hypotonic sucrose solution, water will enter the cells by osmosis in response to the concentration gradient. The osmotic pressure will increase, though not as much as in pure water. If the leaf is placed in a hypertonic sucrose solution, water will leave the cells by osmosis. As a result of the net flow of water out of the cell, the plasma membrane and cell contents shrink away from the cell wall. This process is termed *plasmolysis*. (The plasma membrane is *not* lysed or disrupted.) If the extent and duration of plasmolysis are not too great, there may be no permanent damage. Recovery can occur if the plasmolyzed cell is transferred to a hypotonic solution; water flows into the cell, and *deplasmolysis* occurs. After recovery, water continues to enter by osmosis until equilibrium is reached. You will measure the tonicity of the *Elodea* cytoplasm by placing leaves in progressively higher concentrations of sucrose. The isotonic concentration lies between the highest concentration that prevents plasmolysis and the lowest concentration in which plasmolysis is observed.

PENETRATION OF NONELECTROLYTES

In this project, you will be measuring the rates at which various alcohols penetrate the plasma membrane of *Elodea* leaf cells. Nonelectrolytes generally penetrate by *passive transport*, where the driving force is the concentration gradient across the membrane. The *rate* of penetration is determined by several factors. Generally, the rate of penetration increases with increasing lipid solubility of the permeant. The lipid solubility of a compound can be specified by the ratio of its solubility in an organic solvent to its solubility in water. A frequently used measure of lipid solubility is the *ether:water partition coefficient*, defined as (solubility in diethyl ether)/(solubility in water). Table 9.1 gives the ether:water partition coefficients of the alcohols that will be tested.

TABLE 9.1
Ether:water partition coefficients (solubility in diethyl ether)/(solubility in water) for a series of alcohols. The alcohols are grouped according to the number of carbon atoms and are arranged within each group by increasing number of hydroxyl groups.

Alcohol	Condensed structural formula	Partition coefficient
Methanol	CH_3OH	0.14
Ethanol	CH_3CH_2OH	0.26
Ethylene glycol	$HOCH_2CH_2OH$	0.0053
1-Propanol	$CH_3CH_2CH_2OH$	1.9
Propylene glycol	$CH_3CH(OH)CH_2OH$	0.018
Glycerol	$HOCH_2CH(OH)CH_2OH$	0.00066

The partition coefficients are from R. Collander. (Reprinted with permission from *Acta Chemica Scandinavica*, 1949, Vol. 3, pp. 721–722, Table 1.)

The molecular structure of the permeant is another significant factor in determining its rate of penetration. The condensed structural formulas of the alcohols you will be testing are also given in Table 9.1. An important molecular feature is the presence of hydroxyl groups. Since hydroxyl groups form hydrogen bonds with the water molecules surrounding the cell, a compound with OH groups will have a tendency to remain in the aqueous phase. Molecular size is also a factor; if two compounds have the same lipid solubility but differ in size, it is generally the smaller molecule that enters more rapidly. Small hydrophilic molecules that penetrate rapidly are believed to enter via narrow, protein-lined pores in the plasma membrane.

In some cells, certain molecules penetrate the plasma membrane more rapidly than would be predicted by their molecular structure or lipid solubility. Such *facilitated diffusion*, which is also a passive process, appears to be mediated by specific carrier proteins in the membrane. A number of molecules, especially ions, can enter and leave the cell *against* a concentration gradient. Such *active transport*, as it is termed, requires an expenditure of energy by the cell.

Permeability Study in *Elodea* Leaf Cells.   The present study of permeability with *Elodea* leaf cells is based on the classic experiments of Overton. Each alcohol to which the *Elodea* leaf will be exposed is in an *isotonic* solution of sucrose so that the effect of the alcohol alone can be followed. Rapidly penetrating alcohols will establish equal concentrations almost immediately on both sides of the plasma membrane, and there is no visible effect on the cell. If the alcohol enters slowly, the concentration of the alcohol will be higher outside the cell for a while. During this interval, the water, which can penetrate the plasma membrane very rapidly, leaves the cell, and plasmolysis occurs. As the alcohol penetrates the cell, water reenters by osmosis, and deplasmolysis occurs. The time it takes from the initial exposure to the alcohol solution until deplasmolysis begins is a measure of the time required for penetration.

PROCEDURES

PLASMOLYSIS AND DEPLASMOLYSIS

1. Prepare a wet mount of an *Elodea* leaf in a 0.6 M sucrose solution as follows. Remove a leaf from a healthy sprig, place the leaf on a clean slide with the upper leaf surface facing up, drain the excess water, and add two drops of the sucrose solution and a coverslip.

2. With the green filter removed, examine the preparation under high-dry for plasmolysis. Then, on page 113, draw a plasmolyzed cell, labeling the cell wall, outer margin of the cytoplasm (plasma membrane), inner margin of the cytoplasm where it contacts the vacuole membrane (tonoplast), chloroplasts, and nucleus (if visible).

3. From what part of the cell has most of the water been lost? What is now in the space between the plasma membrane and the cell wall? Enter both answers on page 109 and then clean the slide.

4. Prepare wet mounts of individual *Elodea* leaves in 0.2 M, 0.3 M, 0.4 M, and 0.5 M sucrose solutions, following the same procedure as in step 1. All four wet mounts can be prepared at the same time.

5. Examine each preparation (several fields) under high-dry periodically for 5 min. In each preparation, note whether plasmolysis has occurred and, if so, the extent of plasmolysis. Then answer questions 3–6 on page 109.

6. To observe deplasmolysis, carefully remove the coverslip from a preparation exhibiting plasmolysis, drain the sucrose solution, and add two drops of distilled water. After 1 min, place a coverslip on the preparation. Examine under high-dry for 5 min and then give a brief description of deplasmolysis in item 7 on page 109.

MEASUREMENT OF THE RELATIVE RATES OF PENETRATION OF A SERIES OF ALCOHOLS IN *ELODEA*

The following alcohols are to be tested: methanol, ethanol, ethylene glycol, 1-propanol, propylene glycol, glycerol. The concentration of each alcohol is 0.4 M in a sucrose solution that is approximately isotonic. The alcohol–sucrose solutions should be tested *one at a time* in *Elodea* preparations.

1. Prepare a wet mount of an individual *Elodea* leaf, with the upper surface facing up, in two drops of the alcohol–sucrose solution to be tested. Remember to drain the excess water from the leaf before adding the alcohol–sucrose solution. Record the time that the wet mount is prepared on Data Sheet 9.1

2. Examine the cells under high-dry for the duration and extent of plasmolysis. If plasmolysis does not occur in 5 min, assume that it will not occur, enter *none* for the extent of plasmolysis on Data Sheet 9.1, and proceed testing the next solution. If plasmolysis does occur, record whether it is *slight*, *moderate*, or *severe*, and continue to observe for deplasmolysis.

MEMBRANE PERMEABILITY

3. Watch for the first movement of the cell contents toward the cell wall, using a scale division on the ocular micrometer as a reference point. Record the time that deplasmolysis begins on Data Sheet 9.1. If deplasmolysis does not begin in 10 min, assume that it will not occur and note this on the data sheet.

4. Answer questions 8–14 on pages 109 and 110.

REFERENCES

Davson, H. and Danielli, J. F. 1952. *The Permeability of Natural Membranes*. Cambridge Univ. Press. Reprinted in 1970 by Hafner Press, a division of Macmillan, New York.

Finean, J. B., Coleman, R., and Michell, R. H. 1984. *Membranes and Their Cellular Functions*, 3rd ed. Blackwell Scientific Publications, Oxford, England.

Giese, A. C. 1979. *Cell Physiology*, 5th ed., pp. 371–394. W. B. Saunders, Philadelphia.

Singer, S. J. and Nicolson, G. L. 1972. The fluid mosaic model of the structure of cell membranes. *Science 175*:720–731.

Stadelmann, E. J. 1966. Evaluation of turgidity, plasmolysis, and deplasmolysis of plant cells. In *Methods in Cell Physiology*, Vol. II, pp. 143–216, Prescott, D. M., ed. Academic Press, New York.

EXERCISES AND QUESTIONS

1. During plasmolysis, most of the water is lost from the

_____

2. In a plasmolyzed *Elodea* leaf cell, the space between the plasma membrane and the cell wall contains

_____

3. In which concentration(s) of sucrose is plasmolysis *not* observed?

_____

4. In which concentration(s) of sucrose is plasmolysis observed?

_____

5. Is the extent of plasmolysis the same for all hypertonic concentrations or is the effect progressive?

_____

6. Between which two concentrations does the isotonic concentration of sucrose lie?

_____

_____

7. During deplasmolysis, _____

_____

8. For each alcohol on Data Sheet 9.1, calculate the elapsed time until deplasmolysis began.

9. List the alcohols in the order in which they penetrate the plasma membrane of *Elodea* leaf cells, the most rapid at the top of the list.

_____

_____

_____

_____

_____

_____

10. Describe the relationship between the duration of plasmolysis and the extent of plasmolysis.

_____

_____

11. How do your data from the *Elodea* permeability study demonstrate Overton's rule?

_____

_____

_____

12. Compare the penetration rates of the 2-carbon alcohols, ethanol and ethylene glycol. How can the observed difference be explained on the basis of the ether:water partition coefficients? on the basis of molecular structure?

_____

_____

_____

_____

13. Compare the penetration rates of the 3-carbon alcohols, 1-propanol, propylene glycol, and glycerol. How can the observed differences be explained on the basis of the ether:water partition coefficients? on the basis of molecular structure?

_____

_____

_____

_____

14. Is there a measurable difference in the rate of penetration of the two glycols?

_____ If so, give a reason for the observed difference.

_____

_____

_____

Data Sheet 9.1

Duration and extent of plasmolysis in *Elodea* leaf cells exposed to 0.4 M solutions of various alcohols in an isotonic sucrose solution.

Alcohol	Time wet mount is prepared	Extent of plasmolysis (none, slight, moderate, severe)	Time deplasmolysis begins	Elapsed time (min) until deplasmolysis begins
Methanol				
Ethanol				
Ethylene glycol				
1-Propanol				
Propylene glycol				
Glycerol				

CELL FRACTIONATION

INTRODUCTION

Cell fractionation is used by the cell biologist to investigate the biochemistry and physiology of organelles outside of the complex environment of the intact cell. The basic methodology used to isolate organelles was developed in the 1930s when A. Claude, R. R. Bensley, and N. L. Hoerr isolated mitochondria and S. Granick isolated chloroplasts. The general procedure in all cell fractionation methods consists of disrupting cell boundaries to release the cell contents and then centrifuging to separate the organelles. The purity of each fraction can be confirmed by microscopic examination or with the use of biochemical markers. The primary objective of this project is to isolate chloroplasts, nuclei, and mitochondria from plant tissues by cell fractionation methods. The other objectives are to examine these fractions microscopically and to estimate the *sedimentation coefficient* for the chloroplasts.

HOMOGENIZATION

To prepare tissues for cell fractionation, they must first be minced and suspended in an appropriate medium, often a buffered, isotonic sugar or salt solution. To maintain the integrity of the organelles and enzymes during the fractionation procedure, all solutions and containers must be ice-cold. After being minced, the tissue is ready for *homogenization*, a procedure that disrupts cell boundaries and releases the cell contents, undamaged, into the suspension medium. The suspension of cell constituents is termed the *homogenate*. For plant tissues, homogenization can be carried out with a mortar and pestle (Fig. 10.1), with sand added as an abrasive. In this project, chloroplasts will be isolated from spinach; nuclei and mitochondria will be isolated from cauliflower.

CENTRIFUGATION

Because of differences in size and density, each cell component behaves somewhat differently when subjected to a centrifugal force. The force generated by the centrifuge is expressed as a *relative centrifugal force* (*RCF*) in *g* units, each *g* being the force of the earth's gravity exerted upon a mass. The RCF is a function of the speed of centrifugation in *revolutions per minute* (*rpm*) and the distance of the particle from the axis of rotation. Knowing this distance (x) and the rpm, one can readily determine the RCF from the equation

$$\text{RCF} = 1.119 \times 10^{-5} (\text{rpm})^2 x \qquad (10\text{-}1)$$

where x is taken as the distance (in centimeters) from the axis of rotation to the outer margin of the centrifuge tube. Most *low-speed clinical centrifuges* have a maximum

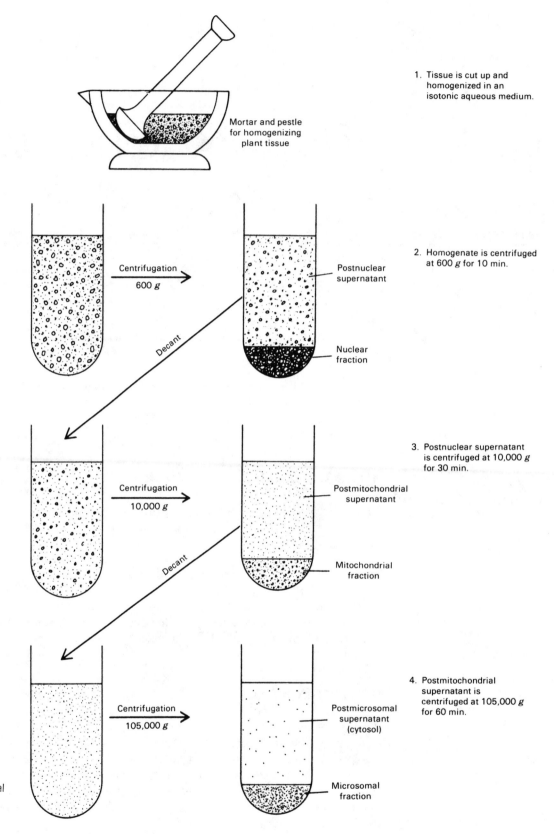

Mortar and pestle
for homogenizing
plant tissue

1. Tissue is cut up and
   homogenized in an
   isotonic aqueous medium.

Centrifugation
600 g

Postnuclear
supernatant

Nuclear
fraction

Decant

2. Homogenate is centrifuged
   at 600 g for 10 min.

Centrifugation
10,000 g

Postmitochondrial
supernatant

Mitochondrial
fraction

Decant

3. Postnuclear supernatant
   is centrifuged at 10,000 g
   for 30 min.

Centrifugation
105,000 g

Postmicrosomal
supernatant
(cytosol)

Microsomal
fraction

4. Postmitochondrial
   supernatant is
   centrifuged at 105,000 g
   for 60 min.

FIGURE 10.1
Cell fractionation
procedure with differential
centrifugation and a
swinging-bucket rotor.

speed of several thousand rpm. A *high-speed centrifuge*, which often has a refrigerated rotor chamber, can operate at speeds up to 25,000 rpm. An *ultracentrifuge* may have a maximum speed in excess of 75,000 rpm. Its rotor chamber is refrigerated and has a partial vacuum.

The two common types of rotors used in centrifuges are the *swinging-bucket rotor* and the *fixed-angle rotor*. In a swinging-bucket rotor there is a mount that allows the centrifuge tubes to pivot from the vertical to the horizontal position during centrifugation. In a fixed-angle rotor, the centrifuge tubes are held at a specific angle, and the material collects at an angle at the bottom of the tubes. A fixed-angle rotor is more efficient than is a swinging-bucket rotor for sedimenting particles. On the other hand, swinging-bucket rotors are much better suited for *density gradient centrifugation*, which is described shortly.

Differential Centrifugation.   In differential centrifugation, the homogenate is centrifuged repeatedly at successively higher speeds, sedimenting progressively smaller particles. The method is illustrated in Figure 10.1. The first centrifugation is at 600 g for 10 minutes, which sediments the *nuclear fraction*. This pellet contains nuclei, intact cells, and tissue debris. For further separation, the *postnuclear supernatant*, or suspended particles and liquid above the nuclear pellet, is transferred to another centrifuge tube and spun at 10,000 g for 30 minutes in a refrigerated, high-speed centrifuge. The sedimented material is termed the *mitochondrial fraction*; it contains mitochondria, microbodies, and, from homogenates of animal tissues, also lysosomes. In studies that require the ribosomes or endoplasmic reticulum, which are in the *microsomal fraction*, the *postmitochondrial supernatant* is centrifuged at very high RCF (105,000 g), using an ultracentrifuge. The *microsomal supernatant* is the soluble phase of the cell, i.e., the cytosol. It should be noted that the isolated fractions are not pure; there is some contamination in each fraction by smaller organelles that initially were at the bottom of the suspension.

Chloroplasts from spinach cells will be isolated using a modification of a method outlined by F. R. Whatley and D. I. Arnon (1962). The spinach is homogenized in a buffered salt solution. After the plant tissue has been ground up in the mortar, the homogenate is filtered through cheesecloth to remove the larger pieces of tissue. The filtrate is first centrifuged at 200 g for 1 minute. The brief centrifugation at low RCF sediments sand, leaf debris, and whole cells. The supernatant is then spun at 1300 g for 5 minutes, sedimenting most of the chloroplasts. Nuclei are also in this fraction but comprise only a small portion of the pellet.

The nuclear and mitochondrial fractions of cauliflower cells will be obtained using a modification of a differential centrifugation procedure outlined by W. D. Bonner (1967). The tissue is homogenized in a buffered mannitol solution. After filtration through cheesecloth, the homogenate is centrifuged as illustrated in Figure 10.1, through step 3.

Density Gradient Centrifugation.   For the best resolution of cell fractions and the least contamination by extraneous organelles, another method, density gradient centrifugation, is used. Here, the homogenate is added to the top of a centrifuge tube that contains a sucrose gradient in which the most concentrated (most dense) sucrose solution is at the bottom of the tube and the least concentrated (least dense) sucrose solution is at the top. Figure 10.2 illustrates the results of *rate separation*, one of several density gradient methods. As the tube spins, the organelles distribute themselves in the sucrose gradient by size, the largest organelles sedimenting most rapidly. With a liver homogenate, for example, after centrifugation at 10,000 g for 20 minutes, the nuclei are in a layer closest to the bottom of the tube, the mitochondria are in a layer above the nuclei, and the microsomal fraction is still higher in the centrifuge tube. The topmost layer contains the cytosol. Note that all the fractions have been separated with a single round of centrifugation. Following centrifugation, the cell fractions are

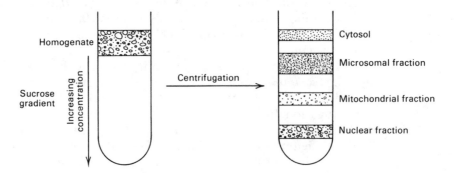

FIGURE 10.2
Cell fractionation with density gradient centrifugation.

Labels (left tube): Homogenate; Sucrose gradient; Increasing concentration

Labels (right tube): Cytosol; Microsomal fraction; Mitochondrial fraction; Nuclear fraction

Centrifugation

removed by puncturing the bottom of the tube and allowing the contents in each layer to drip into a separate test tube.

SEDIMENTATION COEFFICIENT

The *sedimentation coefficient* is a physical constant frequently encountered in biological literature. Many cell structures, especially macromolecules, are often referred to by their sedimentation coefficient. These include proteins, nucleic acids, and small organelles, like the ribosomes. A particle's sedimentation coefficient (s) is determined from its rate of sedimentation under conditions of unit acceleration. Knowing the distance that a particle travels during centrifugation, the duration of the centrifugation, and the rpm, one can calculate s from the equation

$$s = \frac{210 \log (x_2/x_1)}{(\text{rpm})^2 (t_2 - t_1)} \tag{10-2}$$

where x_1 is the distance (in centimeters) from the axis of rotation to the particle at the start of centrifugation, x_2 is the distance from the axis of rotation to the particle at the end of centrifugation, $(t_2 - t_1)$ is the elapsed time (in seconds) of centrifugation, and 210 is a constant. The constant includes a factor that converts the time units in rpm to seconds, so that the units for s are seconds.

An example will illustrate the use of Equation 10-2. Assume that a certain particle initially is at the top of the suspension, 7.6 cm from the axis of rotation ($x_1 = 7.6$ cm). If, after centrifugation at 60,000 rpm for 2.5 hours, the particle is 12.5 cm from the axis of rotation ($x_2 = 12.5$ cm), its sedimentation coefficient is

$$s = \frac{210 \log (12.5/7.6)}{(60,000)^2 (2.5) (60) (60)} = 14 \times 10^{-13} \text{ sec}$$

The s values of most cellular macromolecules lie between 1×10^{-13} sec and 200×10^{-13} sec. For convenience, we let "10^{-13} sec" equal a new unit, the *Svedberg unit* (S). Thus, in the example, the sedimentation coefficient of the hypothetical particle is written as 14S. The sedimentation coefficients have been determined for most cellular macromolecules. The s values are usually determined with an *analytical ultracentrifuge*, in which the sedimenting particles can be observed with a special optical system during the centrifugation.

The factors that determine how rapidly a particle sediments in a centrifugal field are the particle's *radius* and its *effective density*, which is the difference between its density and the density of the liquid through which it is moving. The more important factor is the particle's radius so that, in general, the larger the particle, the greater is its sedimentation coefficient. In fact, knowing the sedimentation coefficient of a macromolecule, one can determine the particle's molecular weight. Not surprisingly, the sedimentation coefficients of the larger organelles (mitochondria, chloroplasts, nuclei) are orders of magnitude greater than are the values obtained for macromolecules.

FIGURE 10.3

Horizontal orientation of a centrifuge tube in a swinging-bucket rotor during centrifugation: x_1 = distance from the axis of rotation to the top of the suspension; x_2 = distance from the axis of rotation to the top of the pellet; A = distance from the bottom of the centrifuge tube to the top of the suspension; and B = distance from the bottom of the centrifuge tube to the top of the pellet. Distances x_1 and x_2 can be obtained after centrifugation by measuring A and B; specifically, x_1 = rotor radius − A and x_2 = rotor radius − B.

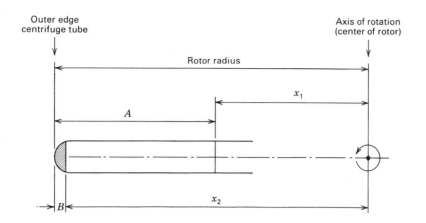

In this project, a sedimentation coefficient for spinach chloroplasts is calculated from the conditions of sedimentation. The chloroplast fraction suspended in a cold, buffered salt solution is centrifuged 5 min in a swinging-bucket rotor. During the centrifugation, a pellet of chloroplasts collects at the bottom of the centrifuge tube. After centrifugation, it is possible to calculate an s value for those chloroplasts that moved from the top of the suspension to the top of the pellet. To calculate s, two measurements are needed: the distance from the axis of rotation at the start of centrifugation (x_1) and the distance from the axis of rotation at the end of centrifugation (x_2). These distances are depicted in Figure 10.3.

MICROSCOPIC EXAMINATION OF THE FRACTIONS

All the isolated fractions will be examined microscopically to identify the organelles present. In addition, the organelles will be measured.

Chloroplasts. Chloroplasts are relatively large organelles, so that some internal detail is discernible with the light microscope. The electron microscope reveals that the chloroplast has two limiting membranes (Fig. 10.4). Internally, there are *grana* (sing., *granum*), which are stacks of flattened membranous sacs. Each flattened sac is called a *thylakoid*. The matrix surrounding the internal membranes is termed the *stroma*. The thylakoids of different grana are connected by an anastomosing network of membranous channels, the *stroma lamellae*.

After the normal morphology of the spinach chloroplasts has been examined

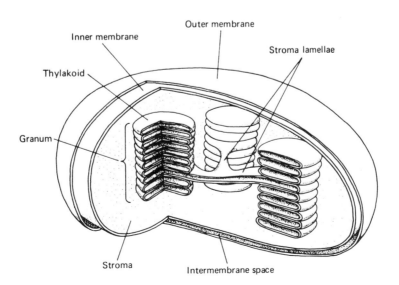

FIGURE 10.4

Three-dimensional representation of a chloroplast as reconstructed from electron micrographs.

microscopically, you will study the effects of two organic compounds on chloroplast morphology. The chloroplasts in two wet mounts will be treated with a saturated aqueous solution of *urea* and a 2% aqueous solution of *Triton X-100*, which is a strong detergent. Urea is known to denature proteins, some of which may thereby be extracted from the chloroplast membranes. Detergents, on the other hand, are known to solubilize the lipid bilayer of biological membranes.

Nuclei. For the observations on nuclei, a stain is needed. In this project, the stain to be used is *lacto-aceto-orcein*, which stains chromatin red. The nucleoli stand out since they do not stain with the orcein. Each nucleolus appears as a prominent, round, clear area.

Mitochondria. Mitochondria are stained with the vital dye *Janus green B*. After the staining, the mitochondria appear blue-green, which is the color of the dye in its oxidized form; in its reduced form, the dye is colorless. Presumably, the Janus green is kept in the oxidized state by the cytochrome oxidase system of mitochondria. Since the mitochondria are very small, you will not be able to resolve any internal structure with the light microscope. A reconstruction based on studies with the electron microscope is presented in Figure 11.1 (p. 131).

PROCEDURES

For all fractionation procedures, you should work in teams of four. For the calculation of the sedimentation coefficient and the microscopic examination of the fractions, work individually.

DETERMINATION OF THE CENTRIFUGATION SPEEDS REQUIRED FOR CHLOROPLAST ISOLATION

1. Measure the *rotor radius*, i.e., the distance (in centimeters) from the center of the rotor to the bottom of the centrifuge tube, positioned horizontally in the swinging-bucket rotor. Enter the value in the table on page 125.
2. Using Equation 10-1, calculate the rpm required for each of the two centrifugations (200 *g* and 1300 *g*) used in obtaining the chloroplast fraction. Enter the values in the table on page 125.
3. Using a centrifuge tachometer, determine the correct settings on the centrifuge's speed control to obtain the required rpm for 200 *g* and required rpm for 1300 *g*.

ISOLATION OF THE CHLOROPLAST FRACTION

1. Weigh out 4 g of fresh spinach leaves from which the major veins have been removed.
2. Cut the leaves into small pieces with scissors and place in a chilled mortar with 15 ml of ice-cold Tris-NaCl buffer and a sprinkling of purified sand. Grind the tissue with a chilled pestle for 2 min.
3. Filter the suspension through four layers of cheesecloth into a chilled 15-ml centrifuge tube. Also wring out the juice into the tube.
4. Centrifuge the filtrate (homogenate) at 200 *g* for 1 min. Make sure that the centrifuge tubes are balanced; the tubes opposite each other should have the same total volume.
5. Decant the supernatant into a clean, chilled centrifuge tube and spin at 1300 *g* for 5 min. Again, be sure that the centrifuge tubes are balanced.

6. After centrifugation, measure distances *A* and *B* (see Fig. 10.3). *A* is the distance (in centimeters) from the bottom of the centrifuge tube to the top of the suspension, and *B* is the distance from the bottom of the centrifuge tube to the top of the pellet. Enter these values on page 125; you will use them to calculate the sedimentation coefficient when all of the microscopic observations on the chloroplasts have been made.

7. Decant and discard the supernatant and then, using a graduated cylinder, add 10 ml of ice-cold Tris-NaCl buffer to the pellet in the centrifuge tube. With a Pasteur pipet, thoroughly resuspend the pellet.

8. Cover the centrifuge tube with Parafilm, invert several times, and place in an ice-water bath.

MICROSCOPIC EXAMINATION OF THE CHLOROPLAST FRACTION

1. With a Pasteur pipet, remove a drop of the chloroplast suspension and prepare a wet mount. Use a small drop so that you will not have to press out excess liquid; pressure might damage the chloroplasts.

2. Examine the chloroplasts under high-dry and oil-immersion, with the green filter removed.

3. Measure the length and width of five chloroplasts. Enter these values and the means on page 127.

4. Observe the morphology of several chloroplasts. Look for some in which the interior is not homogeneous and then answer the questions on page 127.

5. On the blank paper provided at the end of this project (p. 129), draw a chloroplast that shows internal detail, labeling as many structures as possible.

MICROSCOPIC EXAMINATION OF UREA-TREATED CHLOROPLASTS

1. Place a small drop of the chloroplast suspension on a clean slide. Add a small drop of the saturated urea solution and a coverslip. Again, use small drops so that you will not have to press out excess liquid.

2. Examine the chloroplasts under high-dry and oil-immersion for about 5 min and then answer the questions on page 127.

3. Draw a chloroplast that shows the altered morphology.

MICROSCOPIC EXAMINATION OF TRITON-TREATED CHLOROPLASTS

1. Prepare another wet mount, this time using a small drop of the chloroplast suspension and a small drop of the 2% Triton X-100 solution.

2. Examine the preparation under high-dry and oil-immersion and then answer the questions on page 127.

DETERMINATION OF THE SEDIMENTATION COEFFICIENT

1. Determine the sedimentation coefficient for the chloroplasts that initially were at the top of the suspension and, after centrifugation, were at the top of the pellet. To determine *s*, you will have to calculate the distances x_1 and x_2. As can be seen from Figure 10.3, x_1 = rotor radius − *A*, and x_2 = rotor radius − *B*. Enter the values for x_1 and x_2 in the table on page 125.

2. Enter the values for rpm and ($t_2 - t_1$) (in seconds). Then, using Equation 10-2, calculate s, showing all calculations. Enter the values for s and S in the table on page 125.

3. The S values encountered in the literature are corrected so that the sedimentation coefficients are what they would be if the suspending medium were pure water at 20° C. Given that the suspending medium used for the chloroplast isolation in this project is a cold salt solution, comment (on p. 125) on whether the S value obtained is an underestimate or overestimate of the S value corrected for sedimentation in pure water at 20° C.

ISOLATION OF THE NUCLEAR AND MITOCHONDRIAL FRACTIONS

1. Using a single-edge razor blade, remove a total of 20 g of the outer 2-3 mm of the cauliflower surface.

2. Place the tissue in a chilled mortar with 40 ml of ice-cold *mannitol grinding medium* and 5 g of cold purified sand. Grind the tissue with a chilled pestle for 4 min.

3. Filter the suspension through four layers of cheesecloth into a chilled 50-ml centrifuge tube; also wring out the juice into the tube. Allow the filtrate to remain undisturbed for about 2 min, while the sand and debris settle.

4. Carefully decant the supernatant into a fresh, chilled 50-ml centrifuge tube and spin at 600 g for 10 min at 0°-4° C. Make sure that the centrifuge tubes are balanced.

5. Decant the postnuclear supernatant into a clean, chilled centrifuge tube(s) and place the centrifuge tube with the nuclear pellet in an ice-water bath.

6. Centrifuge the postnuclear supernatant at 10,000 g for 30 min at 0°-4° C. Again, be sure that the centrifuge tubes are balanced. During the centrifugation, the nuclear pellet can be examined microscopically, as described below.

7. Decant and discard the postmitochondrial supernatant and add 5.0 ml of the ice-cold mannitol medium* to the mitochondrial pellet.

8. With a spatula, scrape the mitochondrial pellet from the wall of the centrifuge tube and then, with a Pasteur pipet, thoroughly resuspend the sediment in the mannitol medium.

9. Transfer the mitochondrial suspension to a test tube and place in an ice-water bath.

MICROSCOPIC EXAMINATION OF THE NUCLEAR FRACTION

1. With a spatula, remove a tiny amount of the nuclear pellet and smear the material on a clean slide. Immediately, before the smear dries, add several drops of the nuclear stain lacto-aceto-orcein. After 15 sec, add a coverslip and *very gently* press out the excess stain with a paper towel.

2. Examine the preparation under high-dry, with the green filter in place. When you have located a region with nuclei, switch to the oil-immersion objective. For each of five round nuclei, measure the nuclear diameter. Enter the five values and the mean on page 128.

3. On the blank paper provided (p. 129), draw a typical nucleus, labeling each nucleolus.

*CAUTION: *Do not pipet by mouth.*

MICROSCOPIC EXAMINATION OF THE MITOCHONDRIAL FRACTION

1. Place five drops of the mitochondrial suspension in a small test tube. Add five drops of the Janus green stain, swirl the tube, and allow the mixture to remain at room temperature for 10 min.

2. Place a drop of the mixture on a clean slide, add a coverslip, and examine under high-dry and oil-immersion, with the green filter removed. Identify the mitochondria, which appear as very small, blue-green structures.

3. Measure the length of five mitochondria. Enter the five values and the mean on page 128.

4. Observe the morphology of the mitochondria and then answer the question on page 128.

REFERENCES

Bonner, Jr., W. D. 1967. A general method for the preparation of plant mitochondria. In *Methods in Enzymology*, Vol. 10, Estabrook, R. W. and Pullman, M. E., eds., pp. 126–133. Academic Press, New York.

Sheeler, P. 1981. *Centrifugation in Biology and Medical Science*, pp. 10–94. John Wiley, New York.

Whatley, F. R. and Arnon, D. I. 1962. Photosynthetic phosphorylation in plants. In *Methods in Enzymology*, Vol. 6, Colowick, S. P. and Kaplan, N. O., eds., pp. 308–313. Academic Press, New York.

EXERCISES AND QUESTIONS

DETERMINATION OF THE CENTRIFUGATION SPEEDS REQUIRED FOR CHLOROPLAST ISOLATION

RCF	Rotor radius (cm)	Required rpm
200 g		
1300 g		

ISOLATION OF THE CHLOROPLAST FRACTION

$$A = \text{_____} \text{ cm}; B = \text{_____} \text{ cm}$$

DETERMINATION OF THE SEDIMENTATION COEFFICIENT

x_1	x_2	rpm	$(t_2 - t_1)$	s	S

Compared to the S value for chloroplasts sedimented in pure water at 20° C, is the S value obtained in the Tris-NaCl buffer an underestimate or overestimate? Explain.

_____

_____

_____

OBSERVATIONS AND QUESTIONS

MICROSCOPIC EXAMINATION OF THE CHLOROPLAST FRACTION

Chloroplast length and width (µm)

Chloroplast	1	2	3	4	5		
Length	____	____	____	____	____	$\overline{X} =$ ____	µm
Width	____	____	____	____	____	$\overline{X} =$ ____	µm

What is the shape of an individual chloroplast?

_____

What structures are visible within the chloroplasts?

_____

MICROSCOPIC EXAMINATION OF UREA-TREATED CHLOROPLASTS

What morphological changes are occurring in the chloroplasts?

_____

_____

In light of what you know about the action of urea, explain your observations.

_____

_____

MICROSCOPIC EXAMINATION OF TRITON-TREATED CHLOROPLASTS

What morphological changes have occurred in the chloroplasts?

_____

In light of what you know about the action of Triton X-100, explain your observations.

_____

_____

MICROSCOPIC EXAMINATION OF THE NUCLEAR FRACTION

Nuclear diameters (µm)

_____ _____ _____ _____ _____ $\bar{X}$ = _____ µm

MICROSCOPIC EXAMINATION OF THE MITOCHONDRIAL FRACTION

Mitochondrial lengths (µm)

_____ _____ _____ _____ _____ $\bar{X}$ = _____ µm

Do all mitochondria have the same shape? _____ Describe their morphology.

_____

SUCCINATE DEHYDROGENASE ACTIVITY OF MITOCHONDRIA

INTRODUCTION

Contained within the mitochondria is the biochemical machinery for *cellular respiration*, the aerobic process by which sugars, fatty acids, and amino acids are broken down to carbon dioxide and water and their chemical energy captured as *adenosine triphosphate* (*ATP*). A key series of reactions in this aerobic process is the *tricarboxylic acid* (*Krebs*) *cycle*, a complex pathway involving many enzymes and metabolic intermediates. One of the best studied enzymes in the Krebs cycle is *succinate dehydrogenase*, which catalyzes the oxidation of succinate to fumarate. The objective of this project is to measure the rate of the reaction *in vitro* using the mitochondrial fraction isolated from cauliflower cells.

ENZYMES OF THE MITOCHONDRIA

The morphology of mitochondria can vary from almost spherical to fibrillar depending on cell type, but typically they are sausage-shaped organelles, $0.2–1.0\ \mu m$ in diameter and $1–4\ \mu m$ in length. The structural features of a typical mitochondrion are diagrammed in Figure 11.1. There are two membrane systems, an *outer membrane* that separates the mitochondrial contents from the cytoplasm, and a convoluted *inner membrane*, the folds of which are termed *cristae*. The central region surrounding the cristae contains a gel-like *matrix*.

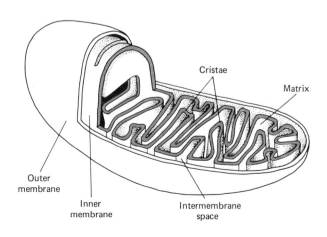

FIGURE 11.1
Three-dimensional representation of a mitochondrion as reconstructed from electron micrographs.

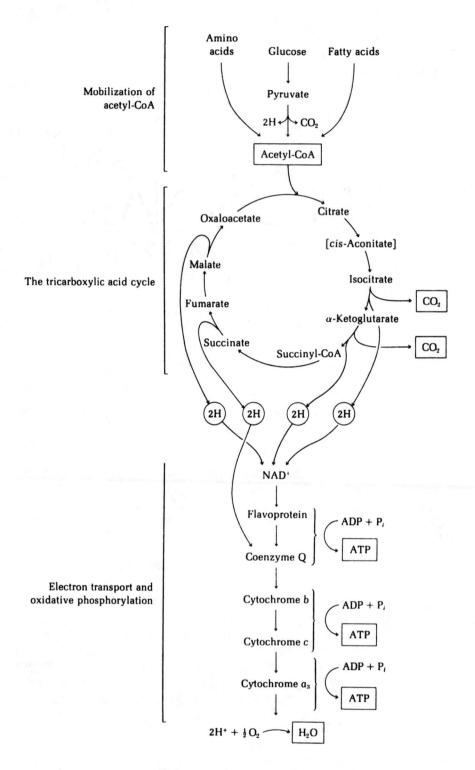

FIGURE 11.2

Summary of the reactions in cellular respiration. (From A. L. Lehninger, 1975, *Biochemistry*, 2nd ed., p. 445, Worth Publishers, Inc., New York.)

The reactions in cellular respiration are shown in Figure 11.2. The reactions from pyruvate through *electron transport* have been localized in the various membranes and compartments of the mitochondrion. The cytochromes and other electron/hydrogen carriers are located in the inner membranes. It is during the process of electron transport that most of the ATP is generated. All but one of the enzymes of the tricarboxylic acid cycle are located in the mitochondrial matrix. The one exception is succinate dehydrogenase and its coenzyme, *flavin adenine dinucleotide* (*FAD*). This enzyme complex, which can be represented by *E-FAD*, is tightly bound to the inner membranes of the mitochondrion. The function of the succinate dehydrogenase–FAD complex is the oxidation of succinate to fumarate:

SUCCINATE DEHYDROGENASE ACTIVITY OF MITOCHONDRIA

$$
\begin{array}{ccc}
\begin{array}{l}
\text{COO}^- \\
| \\
\text{CH}_2 \\
| \\
\text{CH}_2 \\
| \\
\text{COO}^- \\
\text{succinate}
\end{array}
\quad + \quad \text{E-FAD}
\quad \longrightarrow \quad
\begin{array}{l}
\text{COO}^- \\
| \\
\text{CH} \\
\| \\
\text{HC} \\
| \\
\text{COO}^- \\
\text{fumarate}
\end{array}
\quad + \quad \text{E-FADH}_2
\end{array}
$$

As can be seen from this equation, the hydrogens removed from the succinate are carried by FAD. The hydrogens in the reduced complex (E-$FADH_2$) are eventually transferred to coenzyme Q, a component of the electron transport chain (Fig. 11.2).

MEASUREMENT OF SUCCINATE DEHYDROGENASE ACTIVITY

The succinate $\longrightarrow$ fumarate reaction is measured by monitoring the reduction of an *artificial* electron acceptor. To use an artificial electron acceptor, the normal path of electrons in the electron transport chain must be blocked. This is accomplished by adding sodium azide to the reaction mixture. This poison inhibits the transfer of electrons from cytochrome a_3 to the final acceptor, oxygen, so that electrons cannot be passed along by the preceding cytochromes and coenzyme Q. Instead, they can be picked up by an artificial electron acceptor such as *2,6-dichlorophenolindophenol* (*DCIP*). The reduction of DCIP by E-$FADH_2$ can be measured spectrophotometrically since the oxidized form of the dye is blue and the reduced form is colorless. The reaction can be summarized:

$$
\text{E-FADH}_2 \ + \ \text{DCIP}_{ox} \ (\text{blue}) \ \longrightarrow \ \text{E-FAD} \ + \ \text{DCIP}_{re} \ (\text{colorless})
$$

For measurement of succinate dehydrogenase activity in this project, the following reaction mixture is used: assay medium, azide, DCIP, and succinate. The reaction begins when the mitochondrial suspension, which contains the succinate dehydrogenase, is added. The first absorbance reading is taken 5 min later. The change in absorbance at 600 nm is a measure of the extent of the reaction.

ENZYME KINETICS

Enzymes are catalysts, accelerating the rates of reactions so that they occur very rapidly under conditions in the cell. We are interested in factors that affect the rate behavior, or *kinetics*, of enzyme-catalyzed reactions. Pioneering research in enzyme kinetics was reported in 1913 by L. Michaelis and M. L. Menten, who formulated a theory that describes the initial events in enzyme catalysis. The theory assumes that the enzyme (E) and substrate (S) combine reversibly to form an enzyme–substrate (ES) complex. The ES complex then breaks down to form free enzyme and product (P). These reactions are summarized by the equation

$$
E \ + \ S \underset{k_2}{\overset{k_1}{\rightleftharpoons}} ES \xrightarrow{k_3} E \ + \ P
$$

where k_1, k_2, and k_3 are the rate constants. Though the derivation is not presented here, it can be shown that the initial velocity (v_0) of an enzyme-catalyzed reaction is given by the equation

$$
v_0 \ = \ \frac{k_3 \,[E][S]}{\left(\dfrac{k_2 \ + \ k_3}{k_1}\right) \ + \ [S]} \tag{11-1}
$$

We are concerned with the initial velocity and its relationship to enzyme concentration, so we can work with Equation 11-1, which is a form of the

Michaelis–Menten equation. Since the concentration of substrate in the reaction mixture is known and since k_1, k_2, and k_3 are constants, the equation states that the initial velocity is directly proportional to enzyme concentration.

Effect of Enzyme Concentration.   In this project, you will measure the initial velocity of the succinate $\longrightarrow$ fumarate reaction at several concentrations of the enzyme. To vary the concentration of the enzyme in the reaction mixture, you will add different volumes of the resuspended mitochondrial fraction. Since the total volume in each reaction mixture is the same, the relative *volumes* of the added mitochondrial suspension represent the relative *concentrations* of the enzyme in the mixture. Each reaction mixture contains a very high concentration of substrate so that there will still be a relatively high substrate concentration during the first 5 or 10 min of the assay.

The velocity of the reaction will be monitored by measuring the change in absorbance (ΔA_{600}) with time. Recall that ΔA_{600} shows the extent to which the dye DCIP is reduced by E-FADH$_2$. For each of three enzyme concentrations, you will plot ΔA versus time and, from the initial slope of each curve, determine the initial velocity. The initial velocities can then be plotted against enzyme concentration.

Effect of a Competitive Inhibitor.   The velocity of an enzymatic reaction can be altered by the presence of an inhibitor. A *competitive inhibitor* is a substance that competes with the normal substrate for the active site on the enzyme. A classic example of a competitive inhibitor is malonate ($^-OOCCH_2COO^-$), which has a molecular structure similar to that of succinate ($^-OOCCH_2CH_2COO^-$). Though the malonate binds to the succinate dehydrogenase, it cannot be dehydrogenated. Along with the measurement of v_0 at three enzyme concentrations, you will determine the velocity in the presence of malonate.

Controls.   Three controls are required in this experiment. In one control, the reaction mixture contains no azide. In a second control there is no added succinate. The third control contains inactivated enzyme. The enzyme is inactivated by heating the mitochondrial suspension, thereby denaturing the protein.

ISOLATION OF MITOCHONDRIA

The mitochondrial fraction will be isolated from cauliflower cells with the same method of differential centrifugation used in Project 10 (Bonner, 1967). First, the cauliflower tissue must be homogenized in a mortar (Fig. 10.1, p. 116) with a buffered, isotonic mannitol solution and a quantity of sand. After squeezing the homogenate through cheesecloth to remove the larger pieces, you will centrifuge the filtrate at 600 g for 10 min to sediment the nuclear fraction. The supernatant is then centrifuged at 10,000 g for 30 min to sediment the mitochondrial fraction. For all cell fractionation procedures, the solutions and containers must be ice-cold.

PROCEDURES

For all procedures, you should work in teams of four.

ISOLATION OF THE MITOCHONDRIAL FRACTION

1. Using a single-edge razor blade, remove a total of 20 g of the outer 2–3 mm of the cauliflower surface.
2. Place the tissue in a chilled mortar with 40 ml of ice-cold *mannitol grinding medium* and 5 g of cold purified sand. Grind the tissue with a chilled pestle for 4 min.

3. Filter the suspension through four layers of cheesecloth into a chilled 50-ml centrifuge tube; also wring out the juice into the tube.

4. Centrifuge the filtrate at 600 g for 10 min at 0°–4° C. Make sure that the centrifuge tubes are balanced; the tubes opposite each other should have the same total volume.

5. Decant the postnuclear supernatant into a clean, chilled centrifuge tube(s) and spin at 10,000 g for 30 min at 0°–4° C. Again, be sure that the centrifuge tubes are balanced. The nuclear pellet can be discarded.

6. Decant and discard the postmitochondrial supernatant and add 7.0 ml of ice-cold *mannitol assay medium** to the mitochondrial pellet.

7. With a spatula, scrape the mitochondrial pellet from the wall of the centrifuge tube and then with a Pasteur pipet thoroughly resuspend the sediment in the assay medium. It is important that the clumps be completely dispersed.

8. Transfer the mitochondrial suspension to a test tube and place it in an ice-water bath, in which it should be kept during the entire experiment.

MEASUREMENT OF SUCCINATE DEHYDROGENASE ACTIVITY

1. Allow the spectrophotometer to warm up for at least 5 min. Instructions for its use are given in Project 5, pages 42 and 47 (steps 8 and 9 at the bottom). Set the wavelength at 600 nm.

2. Label 10 cuvettes as shown in the table following step 4. Except for the ice-cold mitochondrial suspension, all solutions should be at room temperature.

3. Prepare the mitochondrial suspension for tube 7 by heating a 0.6-ml aliquot in a boiling-water bath for 5 min and then cooling in an ice-water bath.

4. To all cuvettes, add the various solutions given across the top of the table, except for the mitochondrial suspension. First, add the correct volume of assay medium* to all tubes. Then, in the same manner, add the correct volumes of azide,[†] DCIP,* malonate,* and succinate,* as indicated in the table. Cover each cuvette with Parafilm and invert twice to mix the contents.

Tube	Assay Medium*	Azide[†] (0.04 M)	DCIP* (5×10^{-4} M)	Malonate* (0.2 M)	Succinate* (0.2 M)	Mitochondrial suspension*
Blank 1	3.7 ml	0.5 ml	—	—	0.5 ml	0.3 ml
1	3.2 ml	0.5 ml	0.5 ml	—	0.5 ml	0.3 ml
Blank 2	3.1 ml	0.5 ml	—	—	0.5 ml	0.9 ml
2	2.6 ml	0.5 ml	0.5 ml	—	0.5 ml	0.9 ml
Blank 3	3.4 ml	0.5 ml	—	—	0.5 ml	0.6 ml
3	2.9 ml	0.5 ml	0.5 ml	—	0.5 ml	0.6 ml
4	2.7 ml	0.5 ml	0.5 ml	0.2 ml	0.5 ml	0.6 ml
5	3.4 ml	—	0.5 ml	—	0.5 ml	0.6 ml
6	3.4 ml	0.5 ml	0.5 ml	—	—	0.6 ml
7	2.9 ml	0.5 ml	0.5 ml	—	0.5 ml	0.6 ml‡

*CAUTION: *Do not pipet by mouth.*

[†]CAUTION: *Azide is poisonous; do not pipet by mouth.*

‡The mitochondrial suspension to be added to tube 7 will be heated in a boiling-water bath for 5 min.

5. Thoroughly resuspend the mitochondrial suspension with a Pasteur pipet. Then add the correct volume of mitochondrial suspension* to each cuvette, recording the time on Data Sheet 11.1. As soon as the mitochondrial suspension has been

*CAUTION: *Do not pipet by mouth.*

[†]CAUTION: *Azide is poisonous; do not pipet by mouth.*

added, cover the cuvette with Parafilm and invert twice to mix the contents. Remove the Parafilm and set the tube in the test tube rack.

6. When 5 min have elapsed from the time of addition of the mitochondrial suspension to tube 1, adjust the spectrophotometer for blank 1 and take the 5-min absorbance reading for tube 1. Adjust for blank 2 and take the 5-min absorbance reading for tube 2. Adjust for blank 3 and take the 5-min absorbance readings for tubes 3–7. Enter all absorbance readings on Data Sheet 11.1.

7. At 5-min intervals, measure the absorbance of the seven tubes for 35 min. Always remember to adjust for the three blanks as in step 6. Enter the absorbance readings on Data Sheet 11.1.

REFERENCES

Bonner, Jr., W. D. 1967. A general method for the preparation of plant mitochondria. In *Methods in Enzymology*, Vol. 10, Estabrook, R. W. and Pullman, M. E., eds., pp. 126–133. Academic Press, New York.

Lehninger, A. L. 1982. *Principles of Biochemistry*, pp. 207–243, 435–506. Worth, New York.

Tzagoloff, A. 1982. *Mitochondria*. Plenum Press, New York.

EXERCISES AND QUESTIONS

1. On Data Sheet 11.2, enter for each tube the *total change in absorbance* (ΔA) at each time interval. The total change in absorbance is the difference between the 5-min reading for tube 7 and the reading for each tube at the specified time. Why is the 5-min absorbance reading for tube 7 taken as the 0-min reading for all tubes?

_____

_____

2. On Graph 11.1, plot the reaction rates for tubes 1–4. Plot the total change in absorbance (ordinate) versus time (abscissa). Include the origin as a point for all curves, i.e., the 0-min value for ΔA is 0. Draw the best-fit curve with a ruler and French curve template and label each plot.

3. What volume of mitochondrial suspension gives the highest initial velocity?

_____ ml. Explain.

_____

_____

_____

4. On Graph 11.2, plot the initial velocity (ordinate) versus enzyme concentration (abscissa) for tubes 1–3. For the initial velocity, take the total change in absorbance after 5 min (from Graph 11.1) and divide by 5 to obtain the ΔA/min. For the different enzyme concentrations, simply use the three volumes of mitochondrial suspension added: 0.3, 0.6, 0.9. A fourth point should pass through the origin. Draw the best-fit line. Comment on what happens to v_0 when the enzyme concentration is doubled and tripled.

_____

_____

_____

5. How does the initial velocity in the reaction mixture with malonate (tube 4) differ from the initial velocity in the comparable mixture without malonate

(tube 3)? _____
Explain.

_____

_____

_____

6. From the data, what is the evidence that the electron transport system is functioning in the isolated mitochondria?

_____

_____

_____

7. From the data, what is the evidence that there is very little, if any, succinate in the mitochondrial fraction?

_____

_____

_____

NAME ____________ SECTION ________ DATE ________

LAB PARTNERS ____________

Data Sheet 11.1

Absorbance readings (A_{600}) at each time interval.

Tube	Time that mitochondrial suspension was added	Time (min)						
		5	10	15	20	25	30	35
1								
2								
3								
4								
5								
6								
7								

Data Sheet 11.2

Total change in absorbance (ΔA) at each time interval.

Tube	Time (min)						
	5	10	15	20	25	30	35
1							
2							
3							
4							
5							
6							
7							

Graph 11.1
Plot of total change in absorbance (ΔA) versus time (min) for tubes 1–4.

NAME ______________________________ SECTION __________ DATE ______________

LAB PARTNERS ______________________________

Graph 11.2

Plot of initial velocity (v_0) versus relative enzyme concentration for tubes 1–3.

CHROMATOGRAPHY OF PHOTOSYNTHETIC PIGMENTS

INTRODUCTION

The light energy required for photosynthesis is absorbed and trapped by the photosynthetic pigments. The major pigments of photosynthesis are the *chlorophylls*. The two chlorophylls found in green plants are *chlorophyll a* (*chl a*) and *chlorophyll b* (*chl b*). Light energy is also absorbed by *carotenoids* and *phycobilins*, sometimes referred to as *accessory pigments*. Carotenoids occur in all photosynthetic organisms, while phycobilins occur in the red and blue-green algae. The objectives of this project are to separate and to characterize the photosynthetic pigments in spinach leaves. The analysis includes extracting the pigments, separating them with *paper chromatography*, obtaining their absorption spectra, and determining the relative amounts of chlorophylls *a* and *b*.

PHOTOSYNTHETIC PIGMENTS

Chlorophylls. The chlorophylls have a similar molecular structure as shown by the formulas of chlorophyll *a* and chlorophyll *b* in Figure 12.1. Each has a *porphyrin ring* and a long *phytol chain*. The porphyrin ring resembles the prosthetic group of hemoglobin and cytochrome but has a central magnesium atom instead of iron. The alternating double and single bonds of the porphyrin ring make chlorophyll an efficient light-absorbing molecule and determine the general shape of the absorption spectrum. Chlorophylls absorb strongly in the red and blue regions of the visible spectrum and transmit in the green. The phytol chain, which is almost devoid of double bonds, contributes little to the absorption spectrum. As is the case for other compounds, the specific absorption maxima of any chlorphyll depends on the solvent in which it is dissolved.

Carotenoids. There are two classes of carotenoids, the *carotenes* and the *xanthophylls*. Figure 12.2 shows the structural formulas of three common carotenoids in spinach leaves. One is a carotene, β-*carotene*, and the other two are xanthophylls, *lutein* and *violaxanthin*. Another common xanthophyll is *neoxanthin*, not depicted in the figure. All carotenoids have long isoprenoid chains, with alternating double and

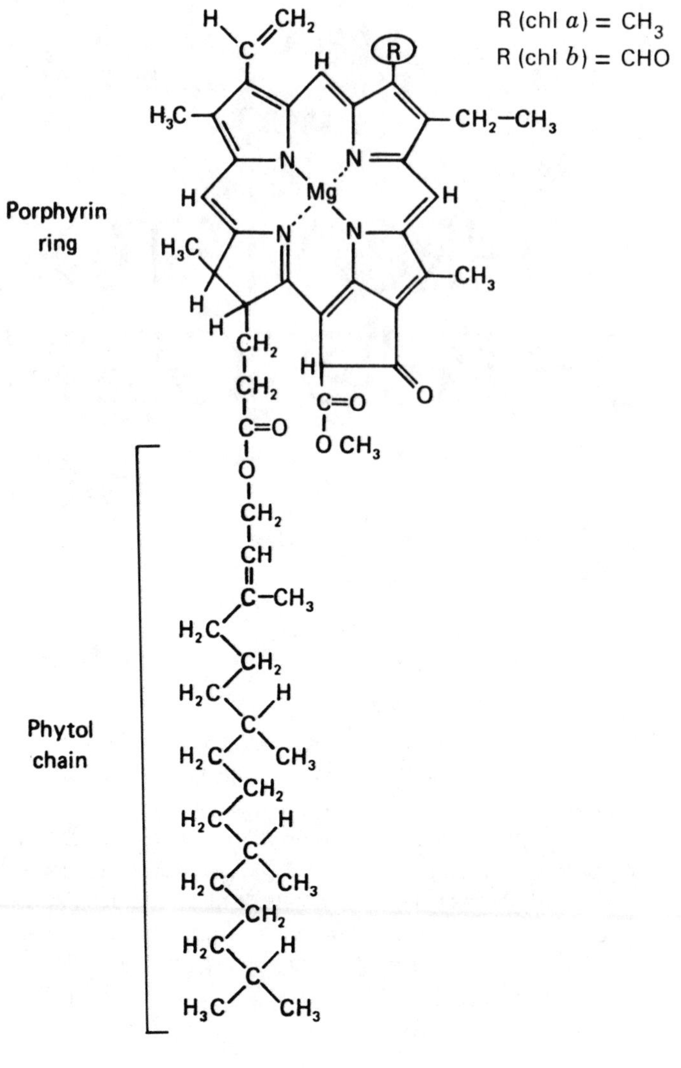

FIGURE 12.1
Structural formulas of chlorophyll *a* (chl *a*) and chlorophyll *b* (chl *b*).

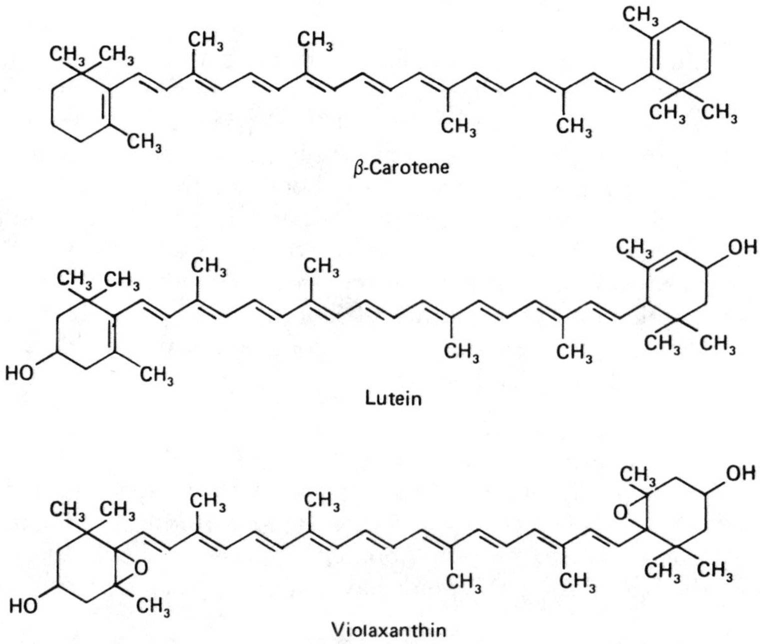

R (chl *a*) = CH₃
R (chl *b*) = CHO

Porphyrin ring

Phytol chain

FIGURE 12.2
Structural formulas of three common carotenoids: β-carotene, lutein, and violaxanthin.

β-Carotene

Lutein

Violaxanthin

CHROMATOGRAPHY OF PHOTOSYNTHETIC PIGMENTS

single bonds. Structurally, the carotenes are composed entirely of carbon and hydrogen, whereas the xanthophylls also contain oxygen. Primarily as a result of their conjugated double bonds, the accessory pigments absorb blue wavelengths of light and transmit yellow and orange light. Some of the absorbed light energy can be transferred to chlorophyll *a*, which participates directly in the conversion of light energy to chemical energy. Since the absorption spectra of the accessory pigments differ from those of the chlorophylls, the effect of accessory pigments is to broaden the range of wavelengths that can be utilized in photosynthesis. Carotenoids also protect the chlorophylls from photooxidation by filtering out harmful wavelengths and by serving preferentially as photooxidizable substrates.

PAPER CHROMATOGRAPHY

Chromatography is a technique used to separate the components of a mixture. There are various types of chromatography (column, paper, thin-layer, gas), but in all cases the separation is achieved by the distribution of components between a fixed or *stationary phase* and a moving or *mobile phase*. In paper chromatography, the components of a mixture are separable into discrete zones on a sheet of filter paper. The mixture is initially spotted or streaked near one end of the paper. If the separated substances are to be extracted later for further analysis, the procedure is called *preparative paper chromatography* (Fig. 12.3). With a capillary tube, the mixture is streaked on the chromatography paper (Fig. 12.3a); enough sample is applied so that

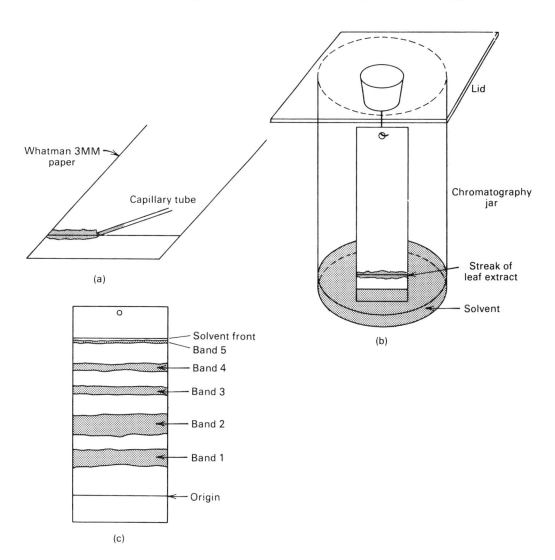

FIGURE 12.3
Preparative paper chromatography. (a) Application of the mixture onto chromatography paper. (b) Chromatography jar with streaked chromatography paper. (c) Chromatogram.

there will be an adequate amount for subsequent extraction and spectrophotometric analysis. For *ascending paper chromatography*, which is depicted in Figure 12.3, the appropriate solvent is added to the bottom of a chromatography jar. The atmosphere in the jar should be saturated with solvent vapor prior to adding the paper. The paper is suspended in the chromatography jar so that the streak is above the level of the solvent (Fig. 12.3b). Then, the solvent moves up the paper by capillary action, past the sample, toward the end of the paper. During this process, termed *development*, the solutes separate and form a trail of discrete bands on the *chromatogram* (Fig. 12.3c).

Mechanism of Separation. In paper chromatography, the stationary phase consists primarily of the water molecules bound to the paper. Chromatography paper contains 6–10% water adsorbed to the cellulose fibers. The solutes in the mixture dissolve to some extent in this bound water. A less significant factor may be the adsorption of the solutes to the cellulose fibers per se. The mobile phase is the moving solvent, in which each solute dissolves to a different extent. Thus, there are two opposing forces. The *driving force*, or moving solvent, tends to carry each solute with it toward the end of the paper. The *resisting force* is due largely to the dissolution of each solute in the cellulose-associated water. It is the interaction of these forces on each component of the mixture that results in the separation of the various solutes.

The degree of separation obtained is largely a function of the relative solubilities of each solute in the stationary (water) and mobile (solvent) phases. Each solute has a characteristic *water:solvent partition coefficient*, defined as (solubility in water phase)/(solubility in solvent phase). During development, each solute is repeatedly partitioned or distributed between the two phases, a process termed *countercurrent distribution*. Figure 12.4 illustrates how countercurrent distribution effects the

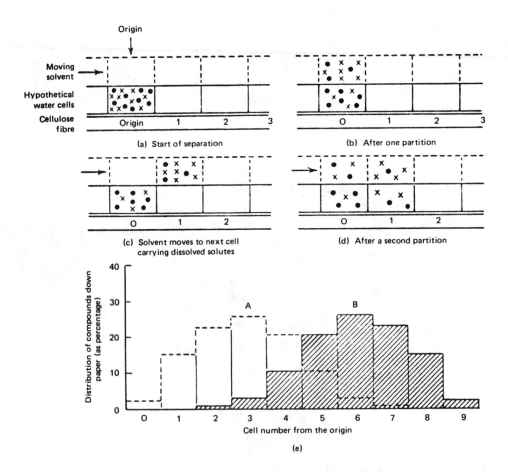

FIGURE 12.4

The process of countercurrent distribution in the separation of solutes A (•) and B (x) by paper chromatography. See text for explanation. (From D. Abbott and R. S. Andrews, 1965, *An Introduction to Chromatography*, p. 4, Houghton Mifflin, Boston. Reproduced by copyright permission of Longman Group Ltd., Harlow Essex, England.)

CHROMATOGRAPHY OF PHOTOSYNTHETIC PIGMENTS

chromatographic separation of a mixture of two solutes, A and B. In this example, A and B have partition coefficients of 2.0 and 0.5, respectively, A being more water-soluble than B. The water associated with the cellulose fibers can be envisioned as occurring in minute "cells." At the *origin*, or point of application, there is initially an equal number of A(•) and B(x) molecules (Fig. 12.4a). As the solvent passes the origin (Fig. 12.4b), each solute distributes itself between the water and solvent according to its partition coefficient (2:1 for A, 1:2 for B). The solvent, now containing some of the mixture, moves over the next cell (Fig. 12.4c). There is now a partition of A and B in cell 1 and a new partition in cell 0 as fresh solvent moves over the origin. As can be seen in Figure 12.4d, there is a tendency for solute B(x) to advance more readily than solute A(•). As the histogram in Figure 12.4e indicates, even 10 such partitions produce a marked separation of A and B. In actuality there are countless cells on a sheet of chromatography paper so that a complete separation can usually be achieved with an appropriate solvent system.

Separation of the components is usually measured by the R_f value. The R_f value is given by the equation

$$R_f = \frac{\text{distance traveled by the solute}}{\text{distance traveled by the solvent from the origin}}$$

For the numerator, the distance is measured from the origin either to the center or to the leading edge of each spot or band. The denominator is the distance from the origin to the solvent front. R_f values can be used to identify the various solutes when the experimental conditions are very carefully controlled.

ANALYSIS OF SPINACH PIGMENTS

The photosynthetic pigments are extracted from spinach by grinding the leaves in acetone. The chromatography paper is then streaked with the spinach extract and suspended in a chromatography jar previously equilibrated with vapors of the solvent, a 9:1 mixture of petroleum ether:acetone. When separation is complete, you will identify the pigment bands by their colors and relative positions on the chromatogram. The major pigments appear in five bands: in order, from the origin to the solvent front, they are chl *b* (olive-green), chl *a* (blue-green), violaxanthin (yellow), lutein (yellow), and, near the leading edge of the solvent front, β-carotene (yellow-orange). Neoxanthin is also present, but its R_f value in this solvent system is close to that of chl *b*, which masks it. Each pigment or pigment group will be eluted from the chromatogram by cutting out each band and soaking the strips of paper in acetone. Violaxanthin and lutein, the two xanthophylls, will be combined and treated as a single group. The absorption spectrum of each pigment or pigment group will then be determined. (The effect of neoxanthin on the absorption spectrum of chl *b* is negligible and can be ignored for this study.)

For the two major pigments, chl *a* and chl *b*, you will perform a quantitative analysis. The absorption coefficients (α) for chl *a* and chl *b* in acetone were determined by G. Mackinney in 1940; for chl *a*, $\alpha_{663} = 80.17$ cm^2/mg, and for chl *b*, $\alpha_{645} = 50.93$ cm^2/mg (663 nm and 645 nm were chosen since they are *absorption peaks* for chl *a* and chl *b*, respectively). Using the Beer–Lambert equation, $A = \alpha cl$ (see Project 5, p. 43), you will determine the concentration (mg/ml) of each chlophyll in the eluted sample. Finally, you will calculate the ratio [chl *a*]/[chl *b*], a value that is characteristic for each plant species.

PROCEDURES

PREPARATION OF THE CHROMATOGRAPHY JAR

For the entire project, you should work in pairs. The solvent mixture is *extremely flammable* so that the chromatography should be carried out in a fume hood. Before handling the chromatography paper, hands should be washed and thoroughly dried. The paper should be handled as little as possible and only by the edges.

1. Cut a strip of Whatman 3MM chromatography paper 7.5 cm wide. The length of the strip should be such that it almost touches the bottom of the chromatography jar when hung from the lid (Fig. 12.3b).
2. In a fume hood, add freshly prepared solvent* (9 vol petroleum ether:1 vol acetone, mixed well) to a height of approximately 2 cm and cover the jar with a lid that is sealed at the rim with Vaseline. Be careful not to get any solvent on the Vaseline. Within 30 min, the atmosphere in the jar will become saturated with solvent vapor.

PREPARATION OF THE LEAF EXTRACT

1. Weigh out 3 g of fresh spinach leaves from which the major veins have been removed.
2. Cut the leaves into small pieces with scissors and then place in a chilled mortar with 15 ml ice-cold acetone and a sprinkling of clean sand. Grind the leaves thoroughly for about 1 min with a chilled pestle.
3. Transfer the liquid and pulp to a 50-ml stoppered test tube. Shake the tube vigorously for 10 sec and then place it in the refrigerator for 10 min. During this time, the white spinach pulp will settle.
4. Using a Pasteur pipet, transfer a portion of the extract, the dark upper layer containing the pigments, to a small stoppered test tube. The pigment extract will be dispensed from this small test tube to minimize evaporation.

PREPARATIVE PAPER CHROMATOGRAPHY

1. With a pencil and ruler, draw a line across the width of the chromatography paper, about 3 cm from the bottom. The extract, which will be streaked on this line, must not be immersed in the solvent.
2. Using a capillary tube, make 10 streak applications of the pigment extract along the line (Fig. 12.3a). The capillary tube is filled by immersing the tip in the extract. The flow from the capillary tube is controlled by finger pressure at the top. Allow each application to air dry before making the next. The final thickness of the streak should be no more than 6–7 mm.
3. Attach the streaked chromatography paper to the hook on the lid of the equilibrated chromatography jar (Fig. 12.3b), being careful not to get any Vaseline on the paper. Allow the chromatogram to develop in the dark or in very dim light for 15–30 min or until there is clean separation of the five bands. Stop the development before the solvent front reaches the end of the paper.
4. Remove the chromatogram from the jar and hold it by the corner until dry. Place the dry chromatogram beneath the chart on Data Sheet 12.1. Next to the chromatogram, mark off in pencil directly on the data sheet the location of the

*CAUTION: *Petroleum ether and acetone are extremely flammable and should be kept away from heat, sparks, or an open flame.*

origin, solvent front, and the approximate center of each band. Also record the color of each band and what pigment it contains. This must be done before elution.

ELUTION AND SPECTROPHOTOMETRY

1. Allow the spectrophotometer to warm up for at least 5 min. Instructions for its use are given in Project 5, pages 42 and 47 (steps 8 and 9 at the bottom). Set the wavelength at 400 nm.

2. Label five cuvettes as follows: chl *b*, chl *a*, xanthophylls, β-carotene, reference blank.

3. Cut out each of the five bands on the chromatogram and place in the appropriate cuvette; place the two xanthophyll bands, violaxanthin and lutein, in the same cuvette. Thus, you will be eluting four pigments or pigment groups: chl *b*, chl *a*, xanthophylls, and β-carotene. You will have to cut the bands into thin strips in order for them to fit in the cuvette.

4. Add 4.0 ml acetone* to each cuvette and cork each tube. Allow the pigments to elute for 5 min, occasionally swirling the tubes. Invert the tubes twice to mix the contents; then remove the paper strips with a pair of forceps, draining the paper against the side of the tube.

5. Measure the absorbance of each sample at the wavelengths listed on Data Sheet 12.2. The spectrophotometer must be adjusted for the reference blank prior to reading the absorbance *at each wavelength*, but all four cuvettes can be read with the one adjustment for the blank at that wavelength. Record the absorbance readings on Data Sheet 12.2

REFERENCES

Goodwin, T. W., ed. 1988 *Plant Pigments*. Academic Press, London.

Govindjee and Govindjee, R. 1974 (Dec.). The absorption of light in photosynthesis. *Sci. Am.* *231*:68–82.

Gregory, R. P. F. 1977. *Biochemistry of Photosynthesis*, 2nd ed. John Wiley, New York.

Sherma, J. and Zweig, G. 1971. *Paper Chromatography and Electrophoresis*, Vol. II. Academic Press, New York.

***CAUTION:** Acetone is toxic; do not pipet by mouth.*

EXERCISES AND QUESTIONS

1. Using the distance markings recorded on Data Sheet 12.1, calculate the R_f value for each pigment. Record the distances and the R_f values in the table on Data Sheet 12.1

2. What do the R_f values indicate about the relative solubilities of the pigments in the water and solvent phases?

_____

_____

_____

_____

3. Explain the relative solubilities of chl b and chl a in the water and solvent phases on the basis of molecular structure.

_____

_____

_____

_____

4. Explain the relative solubilities of the three carotenoids in the water and solvent phases on the basis of molecular structure.

_____

_____

_____

_____

5. On Graph 12.1, plot absorbance (ordinate) versus wavelength (abscissa) for chl b and chl a. Turn the graph paper sideways so that the ordinate is the short axis of the graph paper and the abscissa is the long axis of the graph paper. Each curve should be carefully drawn with a French curve template and then labeled.

6. On Graph 12.2, plot absorbance (ordinate, short axis of graph paper) versus wavelength (abscissa, long axis of graph paper) for the xanthophylls and for β-carotene. Each curve should be carefully drawn with a French curve template and then labeled.

7. At what wavelengths is absorption a maximum for the various pigments?

chl b _____; chl a _____; xanthophylls _____; β-carotene _____

8. From the absorbance of chl *a* at 663 nm and of chl *b* at 645 nm and the Beer–Lambert equation, determine the concentration of chl *a* and chl *b* in the samples.

[chl *a*] = _____ mg/ml; [chl *b*] = _____ mg/ml. Show the calculations.

9. What is the ratio [chl *a*]/[chl *b*] in spinach? _____

Data Sheet 12.1

R_f values for spinach pigments

Pigment	Distance (mm) from origin to solute band	Distance (mm) from origin to solvent front	R_f
β-Carotene			
Lutein			
Violaxanthin			
Chl *a*			
Chl *b*			

Data Sheet 12.2

Absorbance readings for chl *b*, chl *a*, xanthophylls, and β-carotene.

Wavelength (nm)	Chl *b*	Chl *a*	Xanthophylls	β-Carotene
400				
410				
420				
430				
440				
450				
460				
470				
480				
490				
500				
520				
540				
560				
580				
600				
620				
640				
645				
660				
663				
680				
700				

Graph 12.1

Absorption spectra for chl *b* and chl *a*.

Graph 12.2

Absorption spectra for the xanthophylls and β-carotene.

THE HILL REACTION IN ISOLATED CHLOROPLASTS

INTRODUCTION

Photosynthesis is the process by which plants convert light energy to chemical energy. The familiar equation

$$6CO_2 \; + \; 12H_2O \; \xrightarrow[\text{chloroplasts}]{\text{light}} \; C_6H_{12}O_6 \; + \; 6O_2 \; + \; 6H_2O$$

summarizes the process but gives little indication of its complexity. There are actually two sets of reactions, the *light reactions* and the *dark reactions*. As their names imply, one reaction series requires light, while the other does not. The products of the light reactions are molecular oxygen, ATP, and reduced electron carrier. The ATP and reduced carrier are used in the dark reactions to convert carbon dioxide to carbohydrates. The Hill reaction is that important phase of the light reactions in which electrons are transferred from water to an electron acceptor in the presence of light and chloroplasts. The objective of the project is to monitor the rate of the Hill reaction using chloroplasts isolated from spinach. The effect of two inhibitors is also investigated.

LIGHT REACTIONS

All components of the light reactions are located in the thylakoid membranes of the chloroplast (Fig. 10.4, p. 119), while the enzymes of the dark reactions are located in the surrounding stroma. The light reactions of photosynthesis are summarized in Figure 13.1. The figure depicts the flow of electrons in the light reactions, with the height of the various compounds indicating the relative energy level of electrons at each point in the pathway. At the heart of the reaction series are two *photosystems*, designated *photosystem II (PS II)* and *photosystem I (PS I)*. Each photosystem is a light-absorbing assembly of chlorophylls *a* and *b*, carotenoids, cytochromes, and other electron carriers. Each photosystem also contains a special chlorophyll *a* molecule that traps the energy absorbed by the other pigment molecules. This chlorophyll *a* molecule

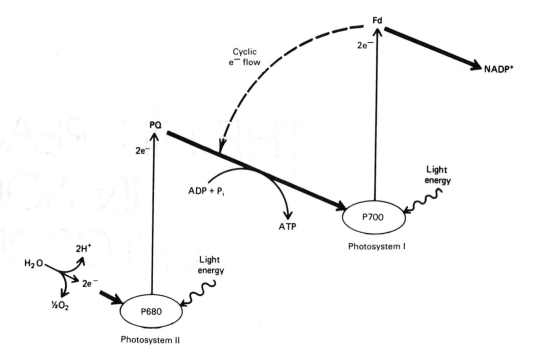

FIGURE 13.1

Path of electrons in the light reactions of photosynthesis. The heavy arrows represent electron transport chains.

absorbs light at a longer wavelength and, hence, at a lower energy level than do the other chlorophyll molecules. In PS II, the energy trap is a chlorophyll *a* molecule that absorbs maximally at 680 nm. In PS I, it is a chlorophyll *a* molecule that absorbs maximally at 700 nm. These special pigment molecules are designated *P680* and *P700*, respectively. The excited electrons in P680 of PS II are transferred to a *plastoquinone* (PQ in Fig. 13.1), and those in P700 of PS I are transferred to *ferredoxin* (Fd in Fig. 13.1), an iron–sulfur protein.

Linking the two photosystems is a major electron transport chain. As the cytochromes and other electron carriers of this chain are successively reduced and oxidized, some of the energy in the electrons is coupled to the synthesis of ATP. The electrons ejected from P700 (in PS I) pass through another electron transport chain and are ultimately passed to *NADP⁺*, the oxidized form of *nicotinamide adenine dinucleotide phosphate*. There is yet one other transfer of electrons, namely, from water to the oxidized P680 of PS II. In a process termed *photolysis*, the details of which are poorly understood, water is split into oxygen, protons, and electrons. These electrons pass via a short electron transport chain to P680.

The flow of electrons from water to NADP⁺ is referred to as the *noncyclic pathway* since the electrons start in one compound (water) and end in another (NADP). The synthesis of ATP via this route is termed *noncyclic photophosphorylation*. The products of the noncyclic pathway of the light reactions are molecular oxygen, reduced electron carrier ($NADPH + H^+$), and ATP. ATP is also produced when electrons take a *cyclic pathway* (Fig. 13.1), which involves PS I and electron carriers in the major electron transport chain. It is a true cyclic flow since the electrons emanate from and return to the same compound: P700. The synthesis of ATP via the cyclic pathway is termed *cyclic photophosphorylation*.

The Hill Reaction. In 1937, Robert Hill showed that isolated chloroplasts can evolve oxygen in the absence of CO_2. This finding was one of the first indications that the source of the electrons in the light reactions is water. In his *in vitro* system, Hill provided an *artificial* electron acceptor. The artificial acceptor intercepts the electrons before they reach PS I, but after the major electron transport chain. The path of electrons from water to the artificial acceptor (A) is, thus,

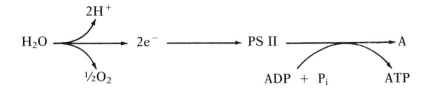

Various dyes can be used as the artificial electron acceptor so that a general equation, known as the *Hill reaction*, can be written

$$H_2O + A \xrightarrow[\text{chloroplasts}]{\text{light}} AH_2 + \tfrac{1}{2}O_2$$

The Hill reaction is formally defined as the photoreduction of an electron acceptor by the hydrogens of water, with the evolution of oxygen. *In vivo*, the final electron acceptor in the light reactions is $NADP^+$.

In isolated chloroplasts, a convenient method for measuring the rate of the Hill reaction is to use as the artificial electron acceptor a dye that changes color as it is reduced. We will use the same dye used in Project 11, 2,6-dichlorophenolindophenol (DCIP), which is blue in its oxidized form and colorless in its reduced form. The change in absorbance, which is measured at 600 nm, will be used to measure the rate of the Hill reaction under a variety of conditions in two experiments. The change in absorbance will be measured at 1-min intervals of exposure to an intense light source. Since the DCIP will begin to revert to its oxidized (blue) state as soon as the chloroplasts (in the reaction vessel) are removed from the light path, it is essential that all absorbance readings be taken as quickly as possible.

EXPERIMENT 1: EFFECT OF INHIBITORS ON THE HILL REACTION

In the first experiment, you will measure the normal rate of the Hill reaction and compare it to the rates in the presence of two inhibitors. One inhibitor is *ammonia*, which is an *uncoupler*, a compound that separates the process of phosphorylation from electron transport. The electron transport system functions, but no ATP is produced. Uncoupling electron transport from ATP synthesis causes an increased electron flow. The dye DCIP can itself be an effective uncoupler. However, at the low concentration used in the present study, there is still *partial* coupling so that the effects of other uncouplers can still be measured. The second inhibitor you will study is *3-(3,4-dichlorophenyl)-1,1-dimethylurea (DCMU)*, an herbicide. DCMU blocks both electron transport *and* phosphorylation by interrupting electron flow at the beginning of the major electron transport chain. The control in Experiment 1 is a reaction mixture kept in the dark.

EXPERIMENT 2: EFFECT OF LIGHT INTENSITY ON THE HILL REACTION

In the second experiment, which you will design yourself, you will determine the effect of light intensity on the rate of the Hill reaction. You will vary the light intensity by placing the reaction vessel at different distances from the light source. Keep in mind that light intensity decreases as the square of the distance from the source.

ISOLATION OF CHLOROPLASTS

The chloroplasts will be obtained from spinach leaves using a modification of a standard fractionation procedure (Whatley and Arnon, 1962), as described in Project 10. You will homogenize the spinach in a mortar with a pestle (Fig. 10.1, p. 116) with a buffered, isotonic salt solution and a small quantity of sand. After squeezing the homogenate through cheesecloth to remove the larger pieces, you will centrifuge the filtrate at 200 g for 1 min, sedimenting debris and whole cells. You will then centrifuge the supernatant at 1300 g for 5 min, sedimenting most of the chloroplasts (and nuclei). For all cell fractionation procedures, the solutions and containers must be ice-cold.

PROCEDURES

For all procedures, you should work in pairs.

ISOLATION OF THE CHLOROPLAST FRACTION

1. Weigh out 4 g of fresh spinach leaves from which the major veins have been removed.
2. Cut the leaves into small pieces with scissors and place in a chilled mortar with 15 ml of ice-cold Tris-NaCl buffer and a sprinkling of purified sand. Grind the tissue with a chilled pestle for 2 min.
3. Filter the suspension through four layers of cheesecloth into a chilled 15-ml conical centrifuge tube. Also wring out the juice into the centrifuge tube.
4. Centrifuge the filtrate at 200 g for 1 min. Make sure that the centrifuge tubes are balanced; the tubes opposite each other should have the same total volume.
5. Decant the supernatant into a clean, chilled centrifuge tube and spin at 1300 g for 5 min. Remember to balance the tubes.
6. Decant and discard the supernatant and then, using a graduated cylinder, add 10 ml of ice-cold Tris-NaCl buffer to the pellet in the centrifuge tube. With a Pasteur pipet, thoroughly resuspend the pellet. To ensure that the chloroplast suspension is thoroughly mixed, cover the centrifuge tube with Parafilm and invert several times.
7. Transfer 4.0 ml of the chloroplast suspension* to a clean, chilled test tube and dilute with 6.0 ml of ice-cold Tris-NaCl buffer*. Cover the test tube, invert several times, and place in an ice-water bath, in which it should remain during the entire experiment. It is the *diluted* chloroplast suspension (hereafter referred to as *chloroplast suspension*) that will be used to measure the Hill reaction.

CAUTION: Do not pipet by mouth.

EXPERIMENT 1: EFFECT OF INHIBITORS ON THE HILL REACTION

1. Allow the spectrophotometer to warm up for at least 5 min. Instructions for its use are given in Project 5, pages 42 and 47 (steps 8 and 9 at the bottom). Set the wavelength at 600 nm.

2. Prepare a water bath as follows. Add 150 ml of water at 20° C to a 250-ml beaker. During illumination, the tubes (cuvettes) are kept in the water bath to keep the reaction temperature fairly constant. Adjust the temperature of the water bath at the beginning of each experimental trial.

3. Place the water bath 25 cm from a lamp with a 100-watt frosted incandescent bulb, but do not turn on the lamp until the start of each experimental trial.

4. Label five cuvettes as shown in the following table. Except for the ice-cold chloroplast suspension, all solutions should be at room temperature.

Tube	Tris-NaCl buffer*	DCIP* $(4 \times 10^{-4}$ M)	Ammonia* (0.01 N)	DCMU* $(10^{-4}$ M)	Distilled water	Chloroplast suspension*
Blank	3.5 ml	—	—	—	1.0 ml	0.5 ml
1[†]	3.5 ml	0.5 ml	—	—	0.5 ml	0.5 ml
2	3.5 ml	0.5 ml	—	—	0.5 ml	0.5 ml
3	3.5 ml	0.5 ml	0.5 ml	—	—	0.5 ml
4	3.5 ml	0.5 ml	—	0.5 ml	—	0.5 ml

*CAUTION: *Do not pipet by mouth.*

†Tube 1 will be wrapped with two layers of aluminum foil so that no light can enter.

5. Wrap tube 1, the nonilluminated control, with two layers of aluminum foil and make a loose-fitting foil cap for the top of the cuvette.

6. Prepare the reference blank and tube 1. Add the solutions* in the sequence given across the top of the table, from left to right. *Before and after* adding the chloroplast suspension, which should be inverted several times just before it is added, cover each cuvette with Parafilm and invert twice to mix the contents. Remove the Parafilm and cover tube 1 with the foil cap. Note the time and set tube 1 aside. An absorbance reading should be taken after 10 min but, meanwhile, proceed with the steps that follow.

7. Adjust the spectrophotometer for the blank.

8. Prepare tube 2. Add the solutions in the same manner used for tube 1 in step 6. Immediately take the 0-min absorbance reading and enter on Data Sheet 13.1.

9. Immediately place tube 2 in the water bath, turn on the lamp, and note the time. After 1 min illumination, remove the tube from the water bath, and *quickly* wipe the surface of the cuvette and measure the absorbance. All absorbance readings must be made as quickly as possible. Enter the absorbance reading on Data Sheet 13.1

10. Return tube 2 to the water bath and take absorbance readings at 1-min intervals of illumination for 10 min. Enter the absorbance readings on Data Sheet 13.1. The spectrophotometer should be adjusted for the blank after every few readings. Remember to take the 10-min absorbance reading for tube 1. Enter it on Data Sheet 13.1. The absorbance should reach zero (or close to it) in 7–10 min.

11. Prepare tube 3, which contains the uncoupler (ammonia). Add the solutions in the same manner used for tube 1 in step 6. Take an absorbance reading immediately and then at 1-min intervals of illumination for 10 min. Remember

CAUTION: *Do not pipet by mouth.*

to adjust the temperature of the water bath to 20° C at the outset and to adjust the spectrophotometer for the blank after every few readings. Enter the absorbance readings on Data Sheet 13.1

12. Prepare tube 4, which contains DCMU. Add the solutions in the same manner used for tube 1 in step 6. Take an absorbance reading immediately and then at 1-min intervals of illumination for 10 min. Remember to adjust the temperature of the water bath to 20° C at the outset and to adjust the spectrophotometer for the blank after every few readings. Enter the absorbance readings on Data Sheet 13.1

EXPERIMENT 2: EFFECT OF LIGHT INTENSITY ON THE HILL REACTION

Design and carry out an experiment to measure the normal rate of the Hill reaction at different light intensities. On Data Sheet 13.1, describe each of the tubes in your experiment, including the contents of each and the distance from the light source. Enter all absorbance readings on Data Sheet 13.1.

REFERENCES

Gregory, R. P. F. 1977. *Biochemistry of Photosynthesis*, 2nd ed. John Wiley, New York.

Hoober, J. K. 1984. *Chloroplasts*. Plenum Press, New York.

Izawa, S. and Good, N. E. 1972. Inhibition of photosynthetic electron transport and photophosphorylation. In *Methods in Enzymology*, Vol. 24, San Pietro, A., ed., pp. 355–377. Academic Press, New York.

Trebst, A. 1972. Measurement of the Hill reaction and photoreduction. In *Methods in Enzymology*, Vol. 24, San Pietro. A., ed., pp. 146–165. Academic Press, New York.

Whatley, F. R. and Arnon, D. I. 1962. Photosynthetic phosphorylation in plants. In *Methods in Enzymology*, Vol. 6, Colowick, S. P. and Kaplan, N. O., eds., pp. 308–313. Academic Press, New York.

EXERCISES AND QUESTIONS

1. On Data Sheet 13.2, enter the *total change in absorbance* (ΔA) at each time interval, i.e., the difference between the initial absorbance reading in each tube and the absorbance reading at the specified time.

2. On Graph 13.1, plot the rate of the Hill reaction for tubes 2–4 (Experiment 1). Plot the total change in absorbance (ordinate) versus time (abscissa). Include the origin as a point for all curves, i.e., the 0-min value for ΔA is assumed to be 0. Draw the best-fit curve with a ruler and French curve template and label each plot.

3. In which tube in Experiment 1 does the reaction proceed most rapidly?

_____ Explain. _____

_____

_____

4. Explain the appearance of the curve for the tube with the DCMU (tube 4).

_____

_____

_____

5. Is there any evidence that the Hill reaction proceeded in tubes 2 and 3 prior to the

time that the 0-min reading was taken? _____

Explain. _____

_____

6. On Graph 13.2, plot the rate of the Hill reaction for the tubes in Experiment 2. Label each plot with the distance from the reaction vessel to the light source.

7. Discuss the relationship between the rate of the Hill reaction and light intensity. Include in your discussion any possible sources of error in Experiment 2.

_____

_____

_____

_____

_____

Data Sheet 13.1
Absorbance readings (A_{600}) at each time interval.

EXPERIMENT 1

Time (min)

Tube	0	1	2	3	4	5	6	7	8	9	10
1	—	—	—	—	—	—	—	—	—	—	
2											
3											
4											

EXPERIMENT 2

Time (min)

Tube	0	1	2	3	4	5	6	7	8	9	10

DESCRIPTION OF TUBES AND EXPERIMENTAL CONDITIONS IN EXPERIMENT 2

_____

_____

_____

_____

_____

_____

Data Sheet 13.2

Total change in absorbance (ΔA) at each time interval.

EXPERIMENT 1

Tube					Time (min)					
	1	2	3	4	5	6	7	8	9	10
2										
3										
4										

EXPERIMENT 2

Tube					Time (min)					
	1	2	3	4	5	6	7	8	9	10

NAME ______________________________ SECTION __________ DATE ____________

LAB PARTNER ______________________________

Graph 13.1

Plot of total change in absorbance (ΔA) versus time (min) for tubes 2–4 (Experiment 1).

NAME ______________________ SECTION __________ DATE ______________

LAB PARTNER ______________________

Graph 13.2

Plot of total change in absorbance (ΔA) versus time (min) for the tubes in Experiment 2.

CELL MOTILITY

INTRODUCTION

It is movement, perhaps more than any other attribute, that distinguishes living from nonliving systems. The locomotion can take many forms. Flagella are whiplike cell appendages that enable spermatozoa and many protozoans to swim. Other protozoans are propelled through their aqueous environment by the coordinated beating of cilia, which are usually shorter than flagella and present in greater numbers on an individual cell. A slower form of cell motility is amoeboid movement, which occurs by a flow of cytoplasm into extensions of the cell surface. It is observed not only in amoebae but in many metazoan cell types, such as leukocytes. Protozoa are ideal subjects for observing and investigating the foregoing modes of locomotion as they are large, easily handled, and readily observed in the living state. The objective of this project is to observe amoeboid movement in an amoeba (*Amoeba proteus* or *Pelomyxa carolinensis*), ciliary motion in *Paramecium caudatum*, and flagellar motion in *Chlamydomonas reinhardi*. An additional objective is to perform an experiment with *Chlamydomonas*, relating flagellar length to the ability of the organism to swim.

AMOEBOID MOVEMENT

The term *amoeboid movement* refers to locomotion resulting from the streaming of cytoplasm into advancing extensions of the cell surface, the *pseudopods*. Amoeboid movement is exhibited not only by amoebae but by various metazoan cells. For example, certain leukocytes move through body tissues and engulf bacteria, and embryonic cells migrate to new regions of the embryo during morphogenesis. This mode of locomotion is most readily observed and studied in amoebae; they are very large (up to several millimeters!), and cytoplasmic changes are obvious with ordinary bright-field or phase-contrast optics.

The familiar freshwater species *Amoeba proteus* (Fig. 14.1) and the larger *Pelomyxa carolinensis* have pseudopods with round ends. The thin peripheral cytoplasm, or *ectoplasm*, is almost devoid of granules and vacuoles. It also has a rather viscous or gelated consistency (*plasmagel*). The internal cytoplasm, or *endoplasm*, is very granular and contains the major organelles. The endoplasm is quite fluid; it is in a solated state (*plasmasol*).

Amoeboid locomotion is accompanied by striking sol–gel transformations at the anterior and posterior ends of the cell (Fig. 14.1). At the anterior end, just behind a thick region of clear cytoplasm termed the *hyaline cap*, the advancing endoplasm spreads laterally and posteriorly, much like the flow of water at the apex of a fountain. As the endoplasm turns posteriorly to become part of the ectoplasm, there is a conversion from plasmasol to plasmagel. At the posterior end, or *uroid*, there is the reverse conversion, from plasmagel to plasmasol.

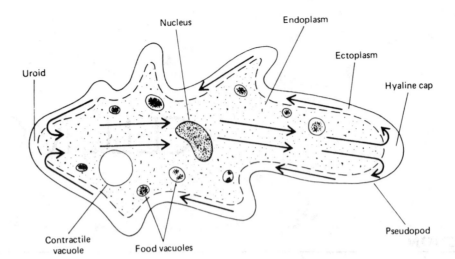

FIGURE 14.1
Amoeba proteus, length approximately 500 μm. Arrows indicate the direction of cytoplasmic flow.

Labels in figure: Uroid, Nucleus, Endoplasm, Ectoplasm, Hyaline cap, Pseudopod, Contractile vacuole, Food vacuoles

While the mechanism responsible for amoeboid movement is poorly understood, it now appears that *actin microfilaments* and possibly *myosin*, as well, are involved in generating the force. Both types of molecules have been extracted from amoebae, and, in electron micrographs, actin-containing microfilaments are visible in the plasmagel. Relating the meshwork of observed microfilaments to amoeboid movement has been perplexing since there is no evidence of an orderly array of filaments, as occurs in striated muscle.

Several models have been proposed to explain amoeboid movement. Two models that address the basic issue of where the contractile force is generated will be examined in this project. In the *tail contraction model*, proposed by S. O. Mast in 1926, the contractile force is generated by the gelated ectoplasm in the uroid region. According to this model, the endoplasm is *pushed* forward into the advancing pseudopod. In 1961, R. D. Allen proposed the *frontal contraction model*, in which the endoplasm is *pulled* forward by a contractile force generated at the tip of the advancing pseudopod. In the first set of observations in this project, you will examine the pattern of cytoplasmic flow in various regions of an amoeba and deduce which of these two models your observations support.

For studying live amoebae, which are rather thick specimens, you will need to use a *Vaseline chamber*. The coverslip in this kind of wet mount is elevated and sealed by a ledge of petroleum jelly placed around its edges. The amoebae will have room to move about and can be observed over an extended period of time without drying out.

CILIARY MOTION AND FLAGELLAR MOTION

There is a basic ultrastructure that is common to both cilia and flagella (Fig. 14.2). The individual cilium or flagellum contains *microtubules*, which are unbranched, hollow cylinders, about 24 nm in diameter, composed entirely of protein. The microtubules are always observed in a characteristic array of nine outer doublets or pairs of microtubules and two central single microtubules. The "9 + 2" core is referred to as the *axoneme*. The axoneme is enclosed by an extension of the plasma membrane so that each cilium and flagellum is really a protruding region of the cytoplasm. At the base of each cilium and flagellum, there is a *basal body*, a barrel-shaped organelle that has the same basic ultrastructure as a centriole. There is compelling evidence that ciliary and flagellar motion results from the doublets sliding past each other (Satir, 1974). Electron micrographs reveal that the doublets on the inner and outer curvature of the bent cilium are the same length but that the doublets on the inner curvature are always closer to the tip of the cilium (Fig. 14.2).

A general distinction between cilia and flagella is that the former are short (5–10 μm) and numerous, whereas the latter are long (up to 150 μm) and sparse. Another

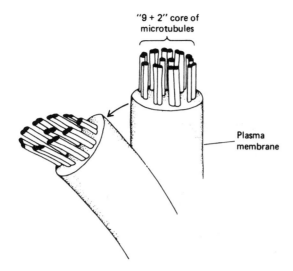

"9 + 2" core of
microtubules

Plasma
membrane

FIGURE 14.2
Organization of
microtubules in cilia and
flagella of eukaryotes.

difference is the nature of their motion. Cilia exhibit an oarlike beat, with alternating power and recovery strokes (Fig. 14.3a). Flagella, on the other hand, usually undulate, with successive waves moving from the base of the flagellum to the tip (Fig. 14.3b). This wave form is observed in the flagella of swimming spermatozoa as well as in many protozoans. Flagella can also exhibit other patterns of movement, as you will observe in *Chlamydomonas*.

Observations on Cilia in *Paramecium*. Ciliary beating will be studied in *Paramecium* (Fig. 14.4). It is probably the most familiar of the ciliates, with its characteristic cigar shape and cilia-covered surface. You will study its swimming motions and the process by which fluid and particles are swept over the surface and into the *buccal cavity* (Fig. 14.4) as the organism feeds. The latter process is similar to the action of the ciliated columnar epithelial cells that line the bronchi and trachea; the beating cilia of these epithelial cells aid in removal of mucous and suspended foreign particles from the respiratory passages.

To make microscopic observations of paramecia, you will have to slow them down by adding yeast cells to the wet mount. The paramecia swim rapidly at first but after 10 minutes congregate around air bubbles and yeast, slowing down as they feed. The yeast cells will have previously been stained with *Congo red* to make them clearly visible.

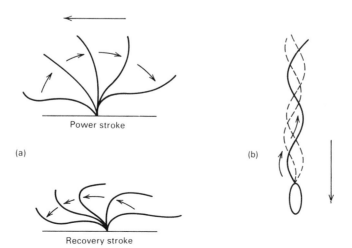

Power stroke

(a)

(b)

Recovery stroke

FIGURE 14.3
Movement of a cilium (a)
and a flagellum (b). The
large arrow indicates the
direction in which the
organism moves.

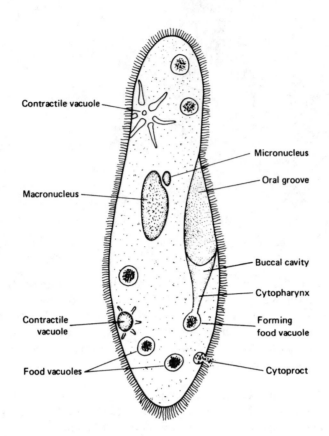

FIGURE 14.4
Paramecium caudatum,
length 170–290 μm.

Contractile vacuole

Micronucleus

Oral groove

Macronucleus

Buccal cavity

Cytopharynx

Contractile
vacuole

Forming
food vacuole

Food vacuoles

Cytoproct

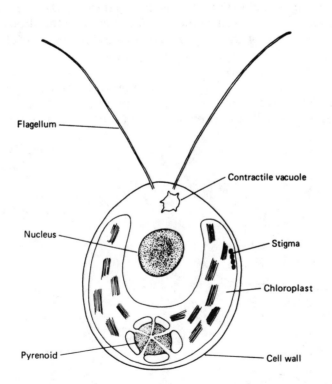

FIGURE 14.5
Chlamydomonas reinhardi,
diameter approximately 10
μm.

Flagellum

Contractile vacuole

Nucleus

Stigma

Chloroplast

Pyrenoid

Cell wall

184

Observations on Flagella in *Chlamydomonas*.   Flagellar motion will be studied in *Chlamydomonas* (Fig. 14.5). *Chlamydomonas* is usually classified as a green alga for it is photosynthetic with a single, large (about 40% of the cell volume), cup-shaped chloroplast and a thin cell wall. At the anterior end of the organism there are two prominent flagella, which function in locomotion. You will study the motility of *Chlamydomonas* in an ordinary wet mount and in a preparation to which a small drop of *Protoslo* (Carolina Biological Supply Co.) has been added. Protoslo is a viscous liquid that is helpful in slowing the organisms for microscopic examination; however, it may alter the normal flagellar motion.

EXPERIMENT ON FLAGELLAR FUNCTION

Chlamydomonas is an excellent subject for studying the relationship between flagellar function and growth. Its suitability for such studies stems from the fact that *Chlamydomonas* can lose or regenerate flagella, depending on culture conditions and experimental manipulations. For example, when grown on solid medium, such as an agar slant, *Chlamydomonas* lose their flagella and grow as a nonmotile form. When returned to liquid medium, the organisms regenerate new flagella and soon become motile. It is this return of motility, which accompanies the flagellar regeneration, that is the focus of the present experiment. Specifically, you will determine the minimum length that the flagella must be for the organisms to swim.

For the experiment on flagellar function, you will be working with normal, flagellated *Chlamydomonas* that have been growing in liquid medium for several days. The first step in the experiment is the removal of flagella, termed *deflagellation*. Deflagellation is accomplished by subjecting the organisms to *pH shock*, a rapid lowering of the pH of the medium. The flagella fall off into the medium, and the deflagellated cells can be recovered by centrifugation. The deflagellated cells are then transferred to fresh liquid medium in which they soon begin regenerating flagella. Accompanying the regeneration is a return of motility, which you will monitor by scoring the percentage of motile organisms in a wet mount. You will also score the percentage of motile organisms in the original (*nondeflagellated*) culture as a control. When normal motility has returned to the deflagellated culture, i.e., when the percentage of motile cells is approximately the same as that in the control culture, you will fix a sample and determine flagellar length at that time.

PROCEDURES

MICROSCOPIC EXAMINATION OF AMOEBAE

1. Using a dissecting microscope, locate the amoebae in the culture dish. With a dropper, carefully transfer one or two organisms to a clean slide. Examine the slide with the dissecting microscope to be sure you have transferred an amoeba and then add a coverslip that has a ledge of petroleum jelly on its edges for support. A neat way of applying the Vaseline is to rub some on the thumb and index finger and then scrape off a small amount with each edge of the coverslip.

2. With the low power objective of a compound microscope, observe the movement of an amoeba over a 10-min period, sketching outlines of the specimen at 2-min intervals. Use the same size scale in each sketch, which should be made on the drawing paper provided (p. 197). Note the changes in the shape and size of the pseudopods. Using an arrow on each drawing, indicate the overall direction of movement of the entire organism.

3. Observe the cytoplasmic streaming in different regions of a pseudopod with low power and high-dry. Continue observing until the organism changes direction.

Closely examine and compare the direction of cytoplasmic streaming in the old and new pseudopods and then answer the questions on page 189.

MICROSCOPIC EXAMINATION OF *PARAMECIUM*

1. On a clean slide, place a drop of the *Paramecium* culture. Dip the tip of a toothpick in the Congo red-stained yeast suspension and then stir the drop with the toothpick. Only a small amount of the yeast suspension should be added; the drop should turn pink, not red. Add a coverslip and carefully seal the edges with Vaseline applied with a clean toothpick.
2. Scan the preparation under low power. While you are waiting for the paramecia to slow down, observe their normal swimming pattern. Identify the anterior and posterior ends. Note that the organism rotates on its own axis. Under high-dry, examine the ciliary motion in various regions of a slow-moving organism and then answer the question on page 189.
3. Examine the yeast cells near the buccal cavity of a slow-moving or stationary specimen. Then answer the question on page 189.
4. Observe the collection of yeast cells in a *food vacuole* at the blind end of the *cytopharynx* (Fig. 14.4). If you continue to observe the food vacuole, you will be able to follow the course of intracellular digestion, specifically, the changes in the size and pH of food vacuoles. Congo red is a pH indicator that is bright orange-red at pH 5 and above, purplish between pH 5 and pH 3, and blue below pH 3.

MICROSCOPIC EXAMINATION OF *CHLAMYDOMONAS*

1. On a clean slide, place a drop of the *Chlamydomonas* culture and a coverslip. Under low power, observe the normal motion of the swimming organisms.
2. Make a fresh wet mount with a drop of the *Chlamydomonas* culture and a tiny drop of Protoslo. Mix with a toothpick and add a coverslip.
3. Examine under low power and high-dry. Close the condenser iris until the flagella are visible. Examine some slowed organisms that are moving forward, i.e., with the flagella at the leading end. Then, on page 189, describe the motion of the flagella.

EXPERIMENT ON FLAGELLAR FUNCTION

Except for the deflagellation, you should work individually. The deflagellation procedure should be carried out by one designated person, since a single deflagellated culture provides enough material for the entire class.

Setting up the Tubes.

1. Place four test tubes in your test tube rack and label them as follows: *Deflagellated/sample 1, Nondeflagellated/sample 1, Deflagellated/sample 2, Nondeflagellated/sample 2.*
2. To each tube, add one drop of Lugol's iodine solution (a fixative and stain) and three drops of culture medium (Medium I of Sager and Granick). These tubes will be used to fix the samples at the outset and at the time when motility has returned.

Deflagellation Procedure.

1. From an actively growing *Chlamydomonas* culture, transfer a 40-ml aliquot to a 100-ml beaker containing a magnetic stirrer. Save the remainder of the nondeflagellated culture as a control.

2. Deflagellate the organisms in the beaker by rapidly lowering the pH of the medium. Monitor the pH with the electrodes of a pH meter. With constant stirring, lower the pH to 4.5 within 30 sec by adding dropwise 0.5 N acetic acid.

3. Wait 30 sec and then restore the pH of the culture to 6.8 by adding dropwise 0.5 N KOH.

4. Pour 10 ml of the deflagellated culture into a 15-ml graduated, conical centrifuge tube. Add 10 ml of water to another centrifuge tube to balance the head and then centrifuge at 1300 *g* for 5 min.

5. Decant and discard the supernatant liquid, which contains the flagella. Using a graduated cylinder, add 10 ml of Medium I to the pellet, which contains the deflagellated cells. Resuspend the cells with a Pasteur pipet.

6. Pour the contents of the centrifuge tube into a 50-ml Erlenmyer flask labeled *Deflagellated*. Also pour a 10-ml aliquot from the original culture into another 50-ml flask labeled *Nondeflagellated*. Place the two cotton-plugged flasks in a gently reciprocating shaker that is illuminated with a fluorescent lamp and proceed immediately to step 1, following.

Determination of the Return of Motility.

1. As soon as possible, remove one or two drops from the deflagellated culture with a Pasteur pipet, add to the test tube labeled *Deflagellated/sample 1*, and swirl the tube to fix the cells rapidly. Using a fresh Pasteur pipet, remove one or two drops from the nondeflagellated culture, add to the test tube labeled *Nondeflagellated/sample 1*, and swirl the tube. Record the time on Data Sheet 14.2. Put these samples aside until the last two samples have been taken. Then all the samples will be examined for flagellar length.

2. Prepare two Vaseline chambers as follows. Place two drops of the deflagellated and nondeflagellated cultures on separate, clean slides, and to each add a coverslip with a ledge of petroleum jelly.

3. Using the low-power objective, examine the "Nondeflagellated" Vaseline chamber for the percentage of motile cells, i.e., the percentage of organisms that are swimming in the microscope field. Count the number of motile and nonmotile cells in three microscope fields, recording the total number of each and the percentage of motile cells for this initial scoring on Data Sheet 14.1. Do the same scoring for the "Deflagellated" Vaseline chamber.

4. Repeat the scoring every 5 min until the percentage of motile cells in the "Deflagellated" Vaseline chamber is approximately the same as that in the "Nondeflagellated" chamber. Immediately after normal motility has returned to the deflagellated cells, fix *sample 2* from the deflagellated and nondeflagellated cultures, using the appropriately labeled test tubes and fresh Pasteur pipets. Record the time on Data Sheet 14.2.

Measurement of Flagellar Length.

1. With a fresh Pasteur pipet, gently resuspend the fixed sample and place a drop on a clean slide. Add a coverslip and examine under low power to locate an area with a concentration of cells.

2. Using a calibrated ocular micrometer with the oil-immersion objective, measure flagellar length in each of 15 cells. Select only cells that have at least one fairly straight flagellum. Since both flagella will be about the same length, just

measure the straighter of the two. Record the measurements for the four samples on Data Sheet 14.2

3. Proceed to Exercises and Questions, page 191.

REFERENCES

Allen, R. D. 1981. Motility. *J. Cell Biol. 91*:148s–155s.

Allen, R. D. and Taylor, D. L. 1975. The molecular basis of amoeboid movement. In *Molecules and Cell Movement*, Inoué, S. and Stephens, R. E., eds., pp. 239–258. Raven Press, New York.

Bray, D. and White, J. G. 1988. Cortical flow in animal cells. *Science 239*:883–888.

Gibbons, I. R. 1981. Cilia and flagella of eukaryotes. *J. Cell Biol. 91*:107s–124s.

Satir, P. 1974 (Oct.). How cilia move. *Sci. Am. 231*:45–52.

OBSERVATIONS AND QUESTIONS

MICROSCOPIC EXAMINATION OF AMOEBAE

When a change in the direction of cytoplasmic streaming occurs, is the change first apparent in the *old* or *new* pseudopod?

_____

In what part of the pseudopod is the movement first observed?

_____

Which of the two hypotheses—tail contraction model or frontal contraction model—do your observations support?

_____

Explain.

_____

_____

_____

_____

MICROSCOPIC EXAMINATION OF *PARAMECIUM*

Do all groups of cilia beat identically? _____ Explain.

_____

_____

Do all yeast cells that enter the buccal cavity get ingested? _____ Explain.

_____

_____

MICROSCOPIC EXAMINATION OF *CHLAMYDOMONAS*

What swimming stroke does the coordinated movement of both flagella resemble?

_____

EXERCISES AND QUESTIONS

EXPERIMENT ON FLAGELLAR FUNCTION

1. From the time you fixed the first samples, how long did it take for normal

motility to return to the deflagellated organisms? _____ min Was the return of

motility sudden or gradual? _____ What does this suggest about
flagellar function in these organisms?

_____

_____

_____

2. For each set of length measurements, calculate the mean, both in ocular
micrometer units and in micrometers. Enter the values on Data Sheet 14.2. What
was the length of the flagella when normal motility returned to the deflagellated

organisms? _____ μm

3. What percentage of the length in controls must be attained before normal

motility returns? _____ %

4. For the time period you studied, what is the mean rate of flagellar regeneration in

Chlamydomonas? _____ μm/min

Data Sheet 14.1

Frequency of motile and nonmotile cells in the deflagellated and nondeflagellated cultures.

Deflagellated

Time (min)	Number of motile cells				Number of nonmotile cells				Total	% Motile cells
Initial										
5										
10										
15										
20										
25										
30										

Nondeflagellated

Time (min)	Number of motile cells				Number of nonmotile cells				Total	% Motile cells
Initial										
5										
10										
15										
20										
25										
30										

Data Sheet 14.2
Flagellar length measurements in the deflagellated and nondeflagellated cultures.

Deflagellated

Sample Time	Flagellar length measurements (ocular micrometer units) in 15 cells															Mean	Mean (μm)
1																	
2																	

Nondeflagellated

Sample Time	Flagellar length measurements (ocular micrometer units) in 15 cells															Mean	Mean (μm)
1																	
2																	

FLAGELLAR REGENERATION IN *CHLAMYDOMONAS*

INTRODUCTION

Flagellar regeneration, as observed in *Chlamydomonas* in the preceding project, is basically a process of microtubule assembly. Each microtubule is a hollow, unbranched cylinder composed of globular protein subunits. Perhaps the most intriguing feature of microtubules is their ability to assemble from globular protein subunits in the cytoplasmic pool. During microtubule assembly, the subunits add onto the ends of existing microtubules, thereby increasing their length. Much of the research on this polymerization process has been carried out on regenerating flagella. The objectives of this project are to study the kinetics of normal flagellar regeneration in *Chlamydomonas* and the effects produced by two chemicals, colchicine and cycloheximide.

THE MICROTUBULE AND ITS ASSEMBLY

All microtubules have a similar ultrastructure. In cross section, the wall of each microtubule is composed of 13 globular subunits (Fig. 15.1). The wall thickness is 5 nm, and the total diameter of the microtubule is about 24 nm. Along the length of the microtubule, which can measure 25 μm or more, the globular subunits form 13 rows, each row termed a *protofilament*.

Ultrastructural analyses of the microtubule have shown that the wall consists of pairs of globular subunits, i.e., *dimers*. Biochemical studies have revealed that there are two types of globular subunits in the dimer, so it is appropriately termed a *heterodimer*. Each heterodimer is composed of one molecule of α-*tubulin* and one molecule of β-*tubulin* (Fig. 15.1). Both are proteins with similar molecular weights (~50,000 daltons) and related amino acid sequences. The heterodimers in the microtubule wall are arranged in a characteristic way, with the α- and β-tubulins alternating along the protofilament and slightly staggered in adjacent protofilaments so as to form a helical array (Fig. 15.1).

In 1972, it was discovered by R. C. Weisenberg that tubulin extracted from cells could assemble into microtubules in a test tube. For such *in vitro* assembly to occur, both *guanosine triphosphate (GTP)* and Mg^{2+} must be present. The process of microtubule assembly is greatly facilitated if *microtubule-associated proteins*, or *MAPs*, are present. MAPs are proteins that remain bound to the tubulin after its extraction from cells. Both assembly and disassembly can occur in the same *in vitro* system. Factors that shift the equilibrium toward disassembly include a relatively high concentration of Ca^{2+} and low temperature.

While the precise conditions that favor microtubule assembly or disassembly in the cell are not known for certain, assembly *in vivo* probably involves the addition of tubulin dimers onto one end of the microtubule. This is depicted in Figure 15.1. The

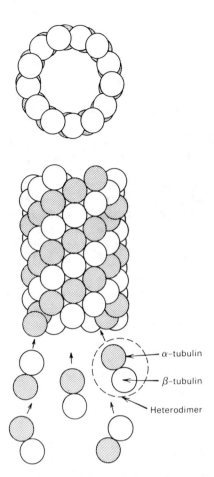

FIGURE 15.1
Diagram showing the arrangement of α-tubulin and β-tubulin in the microtubule and the polymerization of heterodimers at the polymerization end of the microtubule.

α-tubulin

β-tubulin

Heterodimer

figure does not show the process of disassembly, which is believed to occur at the opposite end of the microtubule. Thus, there is a polarity to the microtubule, with polymerization occurring at one end and depolymerization occurring at the opposite end. Flagellar microtubules are relatively stable structures compared, for example, to the microtubules in spindle fibers. In flagella, there appears to be very little depolymerization at the depolymerization end of the microtubules. In cells that are regenerating flagella, it has been shown that the tubulin dimers add on at the *distal* end of the growing flagellum (Rosenbaum, Moulder, and Ringo, 1969).

EXPERIMENT ON FLAGELLAR REGENERATION

The present experiment is based on a study by Rosenbaum, Moulder, and Ringo (1969). It examines the effects of colchicine and cycloheximide on flagellar regeneration in *Chlamydomonas reinhardi*. Colchicine is known to bind to free dimers in the cytoplasmic pool. The colchicine-bound dimers are able to attach to the growing ends of the microtubules but, once attached, prevent any further addition of dimers, whether bound to colchicine or not. Cycloheximide is a potent inhibitor of protein synthesis in eukaryotic cells. In the presence of cycloheximide, there is no longer synthesis of α- or β-tubulin.

The first step in the procedure is the removal of flagella. As in the preceding project, deflagellation is accomplished by rapidly lowering the pH of the *Chlamydomonas* culture. Following the pH shock, the deflagellated cells are divided into three aliquots, each of which is incubated in a different growth medium. One aliquot will be allowed to regenerate flagella in fresh culture medium (Medium I of Sager and Granick). A second aliquot will be placed in Medium I with added colchicine, and the third placed in Medium I with added cycloheximide. From each culture, you will periodically fix samples in Lugol's iodine solution, which also stains the flagella. You will then measure flagellar length in each sample and determine the rates of flagellar regeneration for the three culture conditions. The deflagellated cells incubated in Medium I will serve as a control for measuring the normal rate of regeneration. In addition, the original, nondeflagellated culture will be sampled at the beginning and end of the experiment for a measure of normal flagellar length.

PROCEDURES

EXPERIMENT ON FLAGELLAR REGENERATION

For this experiment, you should work in teams of three, with each member responsible for sampling one of the cultures: Medium I, colchicine, or cycloheximide. The team member responsible for the colchicine culture should also sample and score the nondeflagellated culture. Thus, each team collects data for a complete experiment. The deflagellation procedure should be carried out by one designated team, since a single deflagellated culture provides enough material for the entire class.

Setting up the Tubes.

1. For your particular culture condition, place 10 test tubes in a rack. Note the culture for which you are responsible and label the tubes with the sampling times: 0, 10, 20, 30, 40, 50, 60, 75, 90, and 105 min. The team member responsible for the colchicine culture should also have two additional tubes for sampling the nondeflagellated culture at the start (0 min) and end (105 min) of the experiment.

2. To each tube, add one drop of Lugol's iodine and three drops of Medium I.

Deflagellation Procedure.

1. From an actively growing *Chlamydomonas* culture, transfer a 40-ml aliquot to a 100-ml beaker containing a magnetic stirrer. Save the remainder of the nondeflagellated culture for determination of normal flagellar length.

2. Deflagellate the organisms in the beaker by rapidly lowering the pH of the medium. Monitor the pH with the electrodes of a pH meter. With constant stirring, lower the pH to 4.5 within 30 sec by adding dropwise 0.5 N acetic acid.

3. Wait 30 sec and then restore the pH of the culture to 6.8 by adding dropwise 0.5 N KOH.

4. Pour 30 ml of the deflagellated culture into three 15-ml conical centrifuge tubes, 10 ml per tube, and centrifuge at 1300 g for 5 min.

5. Decant and discard the supernatant liquid, which contains the flagella. To the pellets, which contain the deflagellated cells, add 10 ml of each of the following media, carefully noting which tube gets which medium: Medium I,* Medium I + colchicine (3 mg/ml),† or Medium I + cycloheximide (10 μg/ml).†

6. Resuspend the pellets, using a separate Pasteur pipet for each.

7. Pour the contents of each centrifuge tube into a labeled 50-ml Erlenmyer flask. Also transfer a 10-ml aliquot from the original culture (nondeflagellated cells)* to a labeled 50-ml flask. Place the four cotton-plugged flasks in a gently reciprocating shaker that is illuminated with a fluorescent lamp and proceed immediately to step 1 following.

*CAUTION: *Do not pipet by mouth.*
†CAUTION: *Colchicine and cycloheximide are poisonous; do not pipet by mouth.*

Sampling.

1. As soon as possible, fix the 0-min sample as follows. Using a fresh Pasteur pipet, remove one or two drops from the culture and add to the appropriate test tube. Swirl the tube to fix the cells rapidly. Record the time.

2. Begin measuring flagellar length in the 0-min sample, following the measurement procedure below. Score as many cells as possible between samplings; however, the cells can remain in the fixative and can be scored after all the samples have been taken.

3. Continue sampling every 10 min for the first 60 min and then every 15 min thereafter until the 105-min sample has been fixed. Always be sure that the Pasteur pipet contains no liquid from a previous sampling or use a fresh pipet each time.

Measurement of Flagellar Length.

1. With a fresh Pasteur pipet, gently resuspend the fixed sample and place a drop on a clean slide. Add a coverslip and examine under low power to locate an area with a concentration of cells.

2. Using a calibrated ocular micrometer with the oil-immersion objective, measure flagellar length in each of 15 cells. Select only cells that have at least one fairly straight flagellum. Since both flagella will be about the same length, just measure the straighter of the two. Record the measurements for your culture(s) on Data Sheet 15.1.

REFERENCES

Dustin, P. 1980 (Aug.). Microtubules. *Sci. Am. 243*:66–76.

Lefebvre, P. A. and Rosenbaum, J. L. 1986. Regulation of the synthesis and assembly of ciliary and flagellar proteins during regeneration. *Ann. Rev. Cell Biol. 2*:517–546.

Rosenbaum, J. L., Moulder, L. E., and Ringo, D. L. 1969. Flagellar elongation and shortening in *Chlamydomonas*: The use of cycloheximide and colchicine to study the synthesis and assembly of flagellar proteins. *J. Cell Biol. 41*:600–619.

EXERCISES AND QUESTIONS

1. For each sampling time in your culture(s), calculate the mean flagellar length, both in ocular micrometer units and in micrometers. Enter the values on Data Sheet 15.1

2. On Data Sheet 15.2, record the mean values (in μm) obtained by your team for the three culture conditions and for the nondeflagellated culture.

3. On Data Sheet 15.3, record the combined mean values obtained by the entire class for the three culture conditions and for the nondeflagellated culture.

4. On Graph 15.1, plot flagellar length (ordinate) versus time (abscissa) for the three culture conditions using the combined class data. Draw each curve with a ruler and French curve template and label each plot.

5. In the Medium I culture, do the flagella attain full length by the end of the experiment? _____ In the Medium I culture, would you describe the rate of flagellar regeneration as acceleratory, deceleratory, or constant? Explain.

_____

_____

6. From the data, what is the evidence that colchicine prevents the addition of tubulin dimers onto existing microtubules?

_____

_____

_____

7. How do the early measurements (0–20 min) for the Medium I and cycloheximide cultures compare?

_____

Explain.

_____

_____

_____

8. How do the later measurements for the Medium I and cycloheximide cultures compare?

_____

Explain.

_____

_____

_____

_____

Data Sheet 15.1

Flagellar length measurements for the *Chlamydomonas* culture incubated in _____ (Medium I, colchicine, or cycloheximide) and in the nondeflagellated culture.

Sampling time (min)	Flagellar length measurements (ocular micrometer units)																	Mean	Mean (μm)
0																			
10																			
20																			
30																			
40																			
50																			
60																			
75																			
90																			
105																			

Sampling time (min)	Flagellar length measurements in the nondeflagellated culture																	Mean	Mean (μm)
0																			
105																			

Data Sheet 15.2

Team data for mean flagellar length (μm).

Sampling time (min)	Medium I (N = 15)	Colchicine (N = 15)	Cycloheximide (N = 15)	Nondeflagellated (N = 15)
0				
10				—
20				—
30				—
40				—
50				—
60				—
75				—
90				—
105				

Data Sheet 15.3
Combined class data for mean flagellar length (μm).

Sampling time (min)	Medium I (N =)	Colchicine (N =)	Cycloheximide (N =)	Nondeflagellated (N =)
0				
10				—
20				—
30				—
40				—
50				—
60				—
75				—
90				—
105				

NAME ______________________ SECTION __________ DATE ____________

LAB PARTNERS ______________________

Graph 15.1

Plot of flagellar length versus time for the cultures incubated in Medium I, colchicine, and cycloheximide.

SPERMATOGENESIS IN THE GRASSHOPPER

INTRODUCTION

Meiosis is a basic feature of the life cycle of all sexually reproducing eukaryotes. In a series of two divisions, the diploid number of chromosomes is reduced to the haploid number as the sperm and eggs are formed. Without this reduction, there would be a progressive doubling of the chromosome number in successive generations at fertilization. Besides ensuring that the species' chromosome number remains constant, meiosis generates the bulk of genetic diversity in nature by independent assortment of the chromosomes and crossing over. Considerable knowledge about meiotic events has accrued from studies of spermatogenesis in the grasshopper. One objective of this project is to learn the characteristics of the chromosomes at different stages of meiosis in squash preparations of the grasshopper testis. A second objective is to follow the course of sperm differentiation.

THE TWO MEIOTIC DIVISIONS

The cells that undergo meiosis in the male derive from special cells in the testis, the *spermatogonia*. Prior to meiosis, the spermatogonia increase their numbers by mitosis. After the last mitotic division, DNA synthesis occurs, but there is a change in cell programming; the cells emerge from interphase in prophase of the *first meiotic division* (*M I*). A cell in M I is termed a *primary spermatocyte*. The first meiotic division is sometimes referred to as the *reductional division* since the number of chromosomes is reduced from the diploid to the haploid number as the homologous chromosomes pair and then segregate. The two daughter cells resulting from M I are termed *secondary spermatocytes*. There is no interphase between the two meiotic divisions, and there is no DNA synthesis. In some organisms, there may be a slight uncoiling of the chromosomes and a pause in events, a stage termed *interkinesis*.

The *second meiotic division* (*M II*) is sometimes referred to as the *equational division* because it resembles mitosis. In each secondary spermatocyte, the two chromatids of each chromosome separate and enter the nuclei of the *spermatids*. The four spermatids then differentiate into motile sperm. The transformation, which is termed *spermiogenesis*, involves growth of a flagellum and pronounced changes in the nucleus, Golgi complex, and mitochondria.

While the nuclear events of meiosis are similar in the two sexes, the cytoplasmic events are quite different. During oogenesis there is an unequal distribution of the cytoplasm at both meiotic divisions, resulting in one functional egg and two or three polar bodies, which degenerate.

SPERMATOGENESIS IN THE GRASSHOPPER

Figure 16.1 depicts the essential features of grasshopper spermatogenesis. For simplicity, the figure depicts the cells of a hypothetical organism with 7 chromosomes (three pairs of autosomes and an X chromosome), fewer than would be found in grasshoppers. In many grasshopper species, there are 11 pairs of autosomes and an XX–XO sex-determining mechanism. Thus, there are 24 chromosomes in the female and 23 in the male, usually all acrocentric or telocentric. The X chromosome is *heterochromatic*, meaning that it is condensed and deeply stained at interphase.

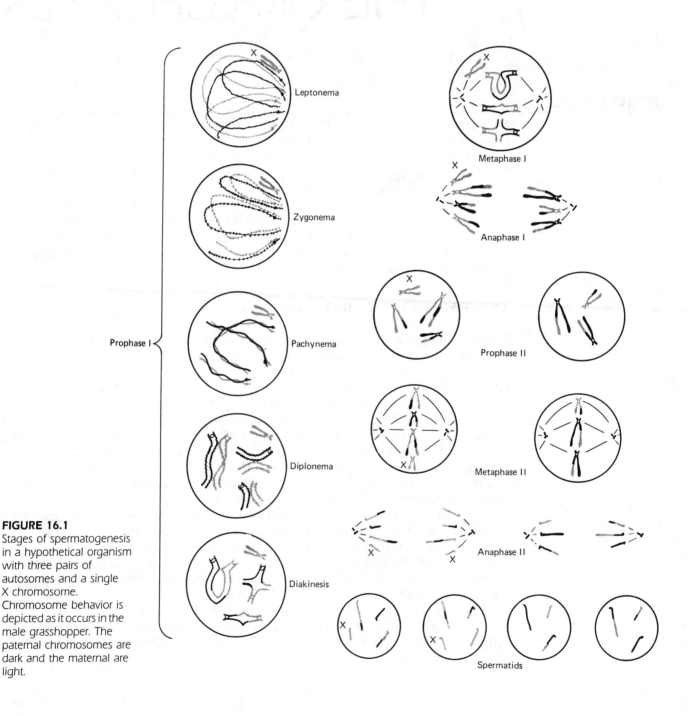

FIGURE 16.1
Stages of spermatogenesis in a hypothetical organism with three pairs of autosomes and a single X chromosome. Chromosome behavior is depicted as it occurs in the male grasshopper. The paternal chromosomes are dark and the maternal are light.

MEIOSIS I

Prophase I.   Prophase of the first meiotic division is much more complex and lasts much longer than mitotic prophase. There are five recognizable stages (Fig. 16.1).

Leptonema.   The first stage of prophase I is *leptonema* (thin-thread stage). Each chromosome is very long and has a distinct pattern of *chromomeres*, localized regions of compacted chromatin. At the centromere there is a block of condensed, deeply stained chromatin, the *centromeric heterochromatin*, which is apparent through the early stages of prophase I. Although DNA replication has taken place, only one subunit is visible in each leptotene chromosome. Frequently, the chromosomes assume a polarized arrangement, with the telomeres (chromosome tips) attached to the nuclear envelope in the vicinity of the centrioles. This *bouquet orientation* may facilitate the initial pairing of the homologues.

Zygonema.   Pairing, or synapsis, occurs in *zygonema* (joined-thread stage). Synapsis can begin anywhere along the chromosomes but probably occurs in a consistent manner for a given chromosome pair. Pairing proceeds in a zipperlike fashion, with the chromomeres and other structural features of the homologues aligning in precise register. All the while, the autosomes are becoming more condensed.

Pachynema.   When pairing is complete, *pachynema* (thick-thread stage) begins. The homologues are much shorter and thicker and are tightly coiled about one another. Each pair of homologues is now referred to as a *bivalent*; the unpaired X is referred to as a *univalent*. It is during pachynema that the exchange of genetic material between homologues, or crossing over, occurs. By late pachynema, the bouquet orientation has disappeared.

Diplonema.   In *diplonema* (double-thread stage), the pairing forces lapse and, as a result, each bivalent opens out. As the name of this stage implies, the double nature of each chromosome is now apparent. Another feature of diplonema is the "fuzzy" appearance of each chromatid. The two chromatids of each homologue are termed *sister chromatids*, and the chromatids of the different homologues of a pair, *nonsister chromatids*. At one or more points along each bivalent there is a cross-shaped connection between nonsister chromatids, termed a *chiasma* (pl., *chiasmata*). Each chiasma represents a point of genetic exchange. Were it not for the presence of at least one chiasma per bivalent, the paired homologues would separate prematurely. Through early diplonema, the centromeres can be identified by the deeply stained centromeric heterochromatin; at later stages of meiosis, the centromere stains similarly to the rest of the chromosome.

Diakinesis.   Diakinesis is the last stage of prophase I. The chromosomes become more condensed and less fuzzy, though the transition between late diplonema and early diakinesis is not always clear. As diplonema and diakinesis progress, the location of some chiasmata may change as a result of repulsion of the homologues and, particularly, of their centromeres. A chiasma that initially was in the middle of an arm may move or slip along the bivalent toward the tip, a phenomenon called *terminalization*. Figure 16.1 depicts several typical bivalent shapes. In late diakinesis, the nucleoli and nuclear envelope disassemble, a spindle forms, and the bivalents move toward the metaphase plate.

Metaphase I.   At *metaphase I*, the bivalents are lined up on the plate. The X univalent, on the other hand, is frequently off the metaphase plate, closer to one pole. The unorthodox location of the X results from its repeated pole-to-pole movements while the autosomes remain lined up on the plate. The arrangement of the homologous pairs is random so that independent assortment occurs. Independent assortment and crossing over together generate the genetic variability brought about by the process of meiosis.

Anaphase I and Telophase I. At *anaphase I* there is synchronous separation of the homologues. The number of arms in each half-bivalent depends on the location of the centromere, which leads the way to the pole. Telocentric and acrocentric chromosomes (as found in most species of grasshopper) have two arms, while metacentrics and submetacentrics have four arms. Following *telophase I*, each of the two nuclei has the haploid number of chromosomes, though each chromosome has two chromatids.

MEIOSIS II

The second meiotic division resembles mitosis with three noteworthy differences: (1) the haploid number of chromosomes is present, (2) the two chromatids of each chromosome are not genetically identical, and (3) the two chromatids are not tightly coiled about one another but are widely separated both at *prophase II* and at *metaphase II*. At metaphase II, the chromosomes are lined up midway between the poles, but in squash preparations they are often observed to be in a circle, an orientation that results when the cell is squashed with the spindle pole at the center. At *anaphase II*, the chromatids of each chromosome separate. As they move to the poles, telocentrics appear as rods, while metacentrics have two arms. At the end of *telophase II*, each spermatid is haploid and genetically unique.

SUPERNUMERARY OR B CHROMOSOMES

In many grasshopper populations, individuals are found with more than the diploid number of chromosomes. Such individuals contain *supernumerary* or *B chromosomes*. B chromosomes, which are usually small and heterochromatic, are additional to the regular chromosomes (*A chromosomes*) and are not homologous with them. Curiously, the B chromosomes have only a slight effect on the phenotype, even when present in multiple doses.

SPERMIOGENESIS

Spermiogenesis is the process by which spermatids differentiate into mature sperm. During sperm development, there are striking changes both in cell shape and in the morphology of various organelles (see review by Phillips, 1970). While young spermatids are nearly spherical with the usual organelles found in most somatic cells, mature sperm are very long and thin and have a unique internal organization. During insect spermiogenesis, certain organelles (centrioles, ribosomes, Golgi complex) are lost, and others (nucleus, mitochondria) undergo dramatic changes in their morphology and physiology. As a result of this differentiation process, animals produce motile sperm that are exquisitely adapted for swimming to the egg, penetrating protective membranes, and effecting fertilization.

In this project, testis material will be stained with *alcoholic HCl–carmine* and fast green FCF. Carmine stains the nucleus red, and fast green stains other parts of the cell green. In squash preparations, young spermatids appear as large, nearly round cells with one or two small clumps of heterochromatin in the nucleus. As the spermatid elongates in the intermediate stages of spermiogenesis, the chromatin is greatly compacted, with one end of the nucleus becoming more tapered than the other. At the posterior end of the nucleus there is a small protuberance that stains bright green. This structure is the *centriole adjunct* and is associated with the pair of centrioles that lie in this region. Flagellar growth is initiated from the more distal of the pair of centrioles, neither of which will be visible in these preparations. The function of the centriole adjunct is not known, but some workers have suggested that it may have a role in securing the flagellum to the sperm head.

Mature sperm are often seen in bundles, which appear as clusters of fine strands. Each sperm cell, which is extremely long and thin, has two distinct regions, the *head* and *tail*. The head is largely the carmine-stained (red) nucleus, which constitutes only a small fraction of the sperm's total length. At the leading, or anterior, end of the sperm head, there is an acrosome, though it is not evident in these preparations. The tail, which stains green with the fast green counterstain, constitutes most of the length of the mature sperm. The tail contains the axoneme of the flagellum along with associated mitochondrial derivatives.

SQUASH PREPARATIONS

Meiosis is best studied in squashes, where all the chromosomes in a cell are present. The squashes will be prepared from testis material previously fixed in 1:3 acetic:ethanol, stained with carmine, and stored in 70% ethanol. Using a dissecting microscope and needles, you will tease apart the *follicles*, the tubular sacs that constitute the testis. Several follicles are placed in a drop of 45% acetic acid on a slide, a coverslip is placed on the tissue, and squashing is carried out with thumb pressure. The 45% acetic acid is an excellent medium for squashing cells. Not only does it soften the tissue, but it also provides a medium of ideal viscosity for dispersing the cells. Since the cells in each follicle exhibit a degree of synchrony, you will have to examine several follicles to observe chromosomes in the various stages of meiosis.

For the study of spermiogenesis, the preparation must be counterstained with fast green. While the counterstained squash may show the stages of meiosis, the appearance of the chromosomes will be better in preparations that are not counterstained. Counterstaining is done when the squash preparation is made permanent. Permanent squash preparations will be made with the quick-freeze method of Conger and Fairchild, discussed in Project 4, page 33.

PROCEDURES

SQUASH PREPARATION

1. With forceps, transfer a stained testis, or part of one, to a Syracuse dish containing 70% ethanol.
2. Using a dissecting microscope and dissecting needles, tease apart several follicles from the testis. Place a drop of 45% acetic acid on a pencil-labeled, alcohol-cleaned slide and place two follicles in the drop. *The tissue must not be allowed to dry out.* Replace the cover on the Syracuse dish.
3. With the dissecting needles, cut each follicle in half and spread out the pieces of tissue. Wait 1 min then add a coverslip.
4. Before squashing, remove excess liquid as follows. Place a paper towel over the slide, hold the towel near one end of the slide, and then move the index finger of the other hand across the slide, pressing lightly. *Be very careful not to move the coverslip laterally*, as this will cause the cells to fold over on themselves, ruining the preparation.
5. With a paper towel held in place on the slide, squash the preparation with the thumb. Use a rolling motion so that the liquid is pressed to the edges of the

coverslip where the towel can absorb it. Thumb pressure should be increased in successive rounds of squashing and not applied all at once. After applying maximum pressure with the thumb, carry out a final round of squashing with the eraser end of a pencil. *Again, be careful not to move the coverslip laterally.*

6. Check the slide for the presence of dividing cells. If meiotic stages are present, the preparation can be made permanent with the procedure that follows. You will need at least two squashes with dividing cells to observe the various stages of meiosis. In addition, you will need a third slide, which need not have any dividing cells, for the study of spermiogenesis. Only the latter slide will be counterstained.

PERMANENT PREPARATIONS AND COUNTERSTAINING

All slides to be made permanent, whether counterstained or not, should be taken through steps 1 and 2. Always use forceps when placing slides in or removing slides from Coplin jars. All staining materials should be kept on Benchkote or paper towels.

1. Place the slide on a flat cake of dry ice* for 5 min.
2. Pry off the coverslip with a single-edge razor blade and *immediately*, before the preparation thaws, immerse the slide in a Coplin jar of 95% ethanol for 1 min. Keep track of which slide surface contains the tissue. To counterstain the squash to be used for the study of spermiogenesis, proceed to steps 3 and 4. To make permanent preparations of the other squashes, proceed directly to step 4.
3. Pass the slide through a second 1-min rinse in 95% ethanol and then immerse it for 30 sec in the counterstain, a 0.5% solution of fast green in 95% ethanol.
4. Rinse the slide in two changes of absolute ethanol, 1 min each, with occasional agitation, and then mount a fresh coverslip with Euparal. Lower the coverslip slowly to minimize air bubbles and then press out the excess liquid with a paper towel.

REFERENCES

Macgregor, H. and Varley, J. 1983. *Working with Animal Chromosomes*, pp. 39–69. John Wiley, New York.

Phillips, D. M. 1970. Insect sperm: Their structure and morphogenesis. *J. Cell Biol. 44*:243–277.

Stephens, R. T. and Bregman, A. A. 1972. The B-chromosome system of the grasshopper *Melanoplus femur-rubrum. Chromosoma 38*:297–311.

Swanson, C. P., Merz, T., and Young, W. J. 1981. *Cytogenetics*, 2nd ed., pp. 200–233. Prentice–Hall, Englewood Cliffs, NJ.

White, M. J. D. 1973. *Animal Cytology and Evolution*, 3rd ed., pp. 148–197; 312–333. Cambridge Univ. Press, London.

*CAUTION: *Dry ice is extremely cold and can injure the skin on contact.*

OBSERVATIONS AND QUESTIONS

1. Scan the non-counterstained preparations under low power, using a green filter. Locate a cell at diplonema and then examine under oil-immersion. Note the morphology of the various bivalents and of the X univalent. How does the staining of the X chromosome compare with that of the autosomes?

   _____

   On page 221, draw two bivalents, labeling the chiasmata. Locate the centromeres by the presence of the deeply stained centromeric heterochromatin. Label them in your drawing.

2. Scan the preparations and locate a cell at metaphase I. Using oil-immersion, note the morphology of the various bivalents and of the X univalent. How does the staining of the X chromosome compare with that of the autosomes?

   _____

   On page 221, draw two bivalents, labeling the centromeres and chiasmata. The centromeres can be identified by the orientation of the bivalents on the metaphase plate.

3. Locate and draw a cell at anaphase I. Label the X chromosome. How did you identify it?

   _____

   _____

4. Locate a group of cells at metaphase II. For each of six cells, count the number of chromosomes and enter these values in the table below.

	Cell					
	1	2	3	4	5	6
Chromosome number						

   How do you explain the variation?

   _____

   _____

5. Are there any supernumerary chromosomes present in your preparation? _____ Explain.

   _____

   _____   _____

6. Scan the counterstained preparation for young spermatids, intermediate stages of spermiogenesis, and mature sperm. In the intermediate stages, is the centriole adjunct found at the tapered end or the blunt end of the head?

On page 223, draw spermatids, intermediate stages of spermiogenesis, and mature sperm, labeling all identifiable structures.

RESTRICTION MAPPING OF PHAGE LAMBDA DNA

INTRODUCTION

Restriction mapping is a widely used method in recombinant DNA research. It is an important adjunct to DNA sequencing methods and, even by itself, reveals a great deal about the molecular architecture of a DNA molecule. The general procedure involves treating a sample of DNA with *restriction enzymes*, which cleave the molecule at specific sites. The method for locating the cleavage sites along the DNA molecule is referred to as restriction mapping. Restriction maps were first used to study the relatively simple genomes of plasmids, viruses, and bacteria. The objective of this project is to obtain a restriction map for the DNA of *bacteriophage lambda* (*phage* λ) using two restriction enzymes.

RESTRICTION MAPPING

Restriction mapping, like many other methods used in molecular genetics research, requires the use of a special class of enzymes, the *restriction endonucleases*, referred to simply as restriction enzymes. Each restriction enzyme recognizes a specific base sequence and cleaves the DNA molecule at every such *recognition sequence*, or *recognition site*. Thus, when a DNA molecule is treated with a particular restriction enzyme, the DNA is cut into a characteristic number of fragments, termed *restriction fragments*. For restriction mapping, the DNA is treated with restriction enzymes individually and in various combinations.

The second step in restriction mapping entails separating the DNA fragments with the technique of *gel electrophoresis*. The enzymatic digests, each containing a number of restriction fragments, are placed in sample wells near one end of the gel. The DNA fragments, which are negatively charged (because of their phosphate groups), migrate toward the anodic (positive) end of the gel at a rate that is a function of molecular weight or length. The distances migrated by the DNA fragments can be used to estimate their lengths.

Once the lengths of the restriction fragments produced by two or more restriction enzymes are known, it is possible to locate the cleavage sites of the enzymes, as the following example illustrates. Consider a DNA molecule containing 50,000 base pairs. The length of a long polynucleotide chain is conveniently given in kilobases (kb), corresponding to the number of kilobase pairs. Thus, the length of this DNA molecule can be written as 50 kb. Now, consider the two restriction enzymes, RE_1 and RE_2. If, in this DNA molecule, there is one recognition sequence for each of these restriction enzymes, treatment with either enzyme would produce two restriction fragments.

Assume that the electropherogram obtained from treatment of this DNA with RE_1 is depicted in Figure 17.1a, lane 1. There are two bands; the band closer to the top of the gel (where the sample was originally loaded) contains fragments that are 35 kb long, whereas the band closer to the bottom of the gel contains fragments that are 15 kb long. In lane 2, we see the pattern of bands that results when the DNA molecule is treated with RE_2. Here, the two fragments produced are 40 kb and 10 kb.

We would like to know in which of the two restriction fragments produced by RE_1 (35 kb or 15 kb) is the recognition sequence for RE_2 located. The answer is obtained by treating the DNA with both enzymes simultaneously and then comparing the lengths of the fragments in the double digest (lane 3) with the lengths of the fragments obtained after treatment with RE_1 alone (lane 1). As can be seen in Figure 17.1a, both of these lanes have a 35-kb fragment, but the 15-kb fragment is missing from lane 3, having been replaced by a 10-kb fragment and a 5-kb fragment. Clearly, then, the cleavage site for RE_2 is located within the 15-kb fragment. A comparison of the fragments produced in the double digest with the fragments produced using RE_2 alone reveals that the cleavage site for RE_1 lies within the 40-kb fragment.

Figure 17.1b shows the restriction map with the cleavage sites of both restriction enzymes indicated. In this simple example, the left and right ends of the DNA molecule (0-kb end and 50-kb end, respectively) are not identified. If the cleavage site in relation to the ends were known for one restriction enzyme, it would be known for the other restriction enzyme when the restriction map was completed.

Restriction Endonucleases.   The various restriction enzymes (more than 175 are known) are isolated from bacteria, where they function to protect the cell from attack by bacteriophages. When phage DNA enters a bacterial cell, it may be fragmented by one or more restriction enzymes in the cell. In fact, the term *restriction* derives from the ability of these enzymes to restrict or limit the host range of bacteriophages. The term *endonuclease* indicates that these enzymes cleave *internal* phosphodiester bonds of polynucleotide chains. The bacterial DNA is unaffected because certain bases have been modified and, hence, are not recognized by the restriction enzymes.

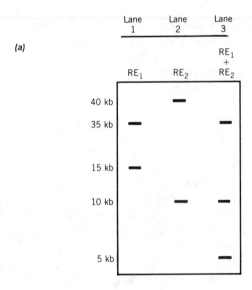

FIGURE 17.1
Electropherogram of restriction fragments (a) and the resultant restriction map (b). See text for explanation.

Restriction enzymes are named for the bacteria from which they are isolated. The name is derived by combining the first letter of the genus with the first two letters of the species. The two restriction enzymes that will be used in this project are *Hin*dIII and *Xho*I. *Hin*dIII is isolated from *Hemophilus influenzae*, and *Xho*I is isolated from *Xanthomonas holicola*. The letter "d" in *Hin*dIII designates the strain, and the Roman numeral indicates whether it was the first, second, or third restriction enzyme to be isolated from that strain.

For molecular genetics research, the most useful restriction enzymes are those designated *type II*. Typically, these enzymes recognize a relatively short base sequence (4–6 base pairs), and then cut each strand within this sequence. For *Hin*dIII, the recognition sequence is

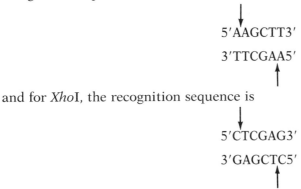

and for *Xho*I, the recognition sequence is

The arrows in each recognition sequence indicate where the cuts are made by the enzyme.

Restriction Mapping of λ DNA.   The structure of phage λ DNA has been thoroughly investigated (Sanger *et al.*, 1972). It contains 48,502 base pairs and has a molecular weight of 30.5 × 10⁶ daltons. Figure 17.2 depicts the restriction map of λ DNA with *Hin*dIII. The figure shows the seven recognition sites and the lengths of the eight restriction fragments. With gel electrophoresis, the six largest restriction fragments separate very well and are frequently used as DNA length standards. Both the 0.6- and 0.1-kb fragments are too small to be detected with the method used in this project.

As an exercise in restriction mapping, you will locate the *Xho*I recognition site in the λ DNA molecule. Lambda DNA contains one recognition sequence for *Xho*I; the lengths of the two restriction fragments are 15.0 kb and 33.5 kb. To determine the location of the *Xho*I recognition site, you will prepare a double digest, in which the λ DNA is treated simultaneously with *Hin*dIII and *Xho*I. From a comparison of the lengths of the restriction fragments present in the *Hin*dIII digest and in the double digest, you will determine the location of the *Xho*I recognition site.

ELECTROPHORESIS OF DNA FRAGMENTS

The gels used for electrophoretic separations of nucleic acids are made of agarose or polyacrylamide. The concentration of the gel can be varied to give different pore sizes; the higher the gel concentration, the smaller the pores. The range of pore sizes determines the suitability of a gel for a particular separation. Gels with large pores

FIGURE 17.2
Restriction map of phage λ DNA with *Hin*dIII. All values are in kilobases (kb). See text for explanation.

are used for separating large polynucleotide chains, whereas those with small pores are used for separating small DNA fragments. This project employs a 1% agarose gel, which is suitable for separating DNA fragments between 0.5 kb and ~6 kb. So that the electrophoretic separation can be completed in a reasonable length of time, the gel must be small, about 90 mm in length. Use of a small gel permits rapid separation but reduces resolution of bands containing large DNA fragments.

The mechanism by which DNA fragments separate during gel electrophoresis has been the subject of considerable research. Unlike proteins, which are separated mainly as a result of differences in electrical charge, nucleic acids are separated on the basis of differences in size. Since every nucleotide contains one negatively charged phosphate, all polynucleotide chains have the same charge-to-mass ratio, and all migrate toward the anode. What appears to be a critical factor in their separation is the *molecular-sieving action* of the porous gel; the smaller the polynucleotide, the more rapidly it weaves its way through the gel matrix. Thus, the smaller a fragment, the higher is its electrophoretic mobility and the farther is its migration through the support matrix in a given period.

DETERMINING THE LENGTHS OF RESTRICTION FRAGMENTS

The length of a DNA fragment can be estimated from the distance it migrates during electrophoresis. It has been determined that the electrophoretic mobility of a DNA fragment is inversely proportional to the logarithm of its molecular weight or, simply, to the logarithm of its length in kilobases. The length of a restriction fragment is determined from a *standard curve*. This is a plot of log kb (ordinate) versus distance migrated (abscissa). To obtain a standard curve, DNA fragments of known lengths are used. In this project, the DNA length standards are the restriction fragments produced when λ DNA is treated with *Hin*dIII. The lengths of the *Hin*dIII fragments are given on the restriction map of λ in Figure 17.2. With the gel used in this project, the plot will be linear for DNA fragments up to 4–5 kb in length and slightly curved for larger fragments.

PREPARING THE SAMPLES

The digests will be prepared by adding the restriction enzymes and reaction buffer (for appropriate ions and pH) to aliquots of λ DNA in micro test tubes. All solutions are added with an *adjustable micropipetter*. Volumes are conveniently given in *microliters* (µl) (1 µl = 10^{-6} l = 10^{-3}ml). In this project, there are four samples: untreated λ DNA, λ DNA + *Xho*I, λ DNA + *Hin*dIII, λ DNA + *Xho*I + *Hin*dIII.

The samples are incubated at 37° C for 30 min, after which the enzymatic reactions are halted by placing the tubes at 65° C and by adding a *"stop" solution*. The "stop" solution contains sodium dodecyl sulfate, which denatures the enzymes, and bromphenol blue, which is a *tracking dye*. The tracking dye moves ahead of all but the smallest DNA fragments, migrating at about the same rate as fragments that are 0.5 kb long. Thus, by observing the tracking dye, the experimenter knows the location of the leading edge of the restriction fragments at any time during the course of electrophoresis. The "stop" solution also contains glycerol, which is required for proper loading of the sample in the agarose gel.

LOADING AND ELECTROPHORESING THE SAMPLES

You will be provided with a previously poured, 3-mm-thick agarose gel ready for loading within the electrophoresis chamber. The chamber used for the electrophoresis of DNA is similar to that used for the separation of proteins (Fig. 6.4, p. 66). There are two buffer reservoirs with a support deck between them for the agarose gel slab. In a horizontal gel apparatus, the gel must be "submerged." If there were no liquid covering the upper surface, the gel would dry out and there would be an uneven voltage gradient through the gel. Since the sample wells are filled with running buffer, a sample can be loaded only if it is denser than the aqueous buffer; i.e., the sample must sink to the bottom of the well. The presence of glycerol in the samples renders them dense enough to be properly loaded. Each sample is loaded in the wells of the agarose gel with the adjustable micropipetter.

The samples are electrophoresed at 60 volts. Electrophoresis is halted when the tracking dye has migrated about 60 mm. The time required for such migration is 1½ to 1¾ hr. *Because of the potential for electrical shock, extreme caution must be exercised; never touch the electrophoresis chamber or any of the electrical connections while the power is turned on.*

STAINING THE DNA BANDS

The bands in the agarose gel will be stained with a thiazolium dye, the common name of which is *Stains-all*. After electrophoresis, the gel is stained overnight in a 0.005% solution of Stains-all in 50% formamide. Since the stain is bleached by light, the staining dish must be wrapped in aluminum foil. The gel is then destained by rinsing with water. Destaining improves the contrast between the stained bands and the surrounding gel. The result is a series of blue bands, each containing DNA fragments of the same length.

PROCEDURES

For all procedures, you should work in teams of four. Each team will prepare and electrophorese four samples: λ DNA, λ DNA + *Xho*I, λ DNA + *Hin*dIII, λ DNA + *Xho*I + *Hin*dIII. Each team will have its own micropipetter, micro test tubes, electrophoresis chamber with agarose gel slab covered with running buffer, and power supply.

SAMPLE PREPARATION

All solutions should be kept cold until dispensed. Your instructor will demonstrate the use of the micropipetter.

1. Label four micro test tubes as shown in the following table.

Tube	Reaction buffer (10 ×)	Distilled water	λ DNA	*Xho*I	*Hin*dIII
1	10 μl	80 μl	10 μl	—	—
2	10 μl	70 μl	10 μl	10 μl	—
3	10 μl	70 μl	10 μl	—	10 μl
4	10 μl	60 μl	10 μl	10 μl	10 μl

2. Add the solutions in the sequence given across the top of the table, from left to right. After each solution is dispensed, gently flick the tube several times to mix the contents. Add each small drop near the bottom of the tube but do not let the pipet tip touch the liquid in the bottom of the tube. (If the drop remains on the side of the tube, you will have to cap it and flick more vigorously to mix the contents.) Be sure to use a fresh pipet tip for each reagent.
3. Place the tubes in the micro test tube block (holder) at 37° C for 30 min.

4. At the end of the incubation, add 10 μl of the "stop" solution to each tube. To mix the contents completely (as indicated by a *uniform* blue color), cap each tube and then swirl the contents.

5. Uncap the tubes and place them in the micro test tube block at 65°C for 5 min.

6. At the end of the incubation, remove the tubes from the block and load the samples.

LOADING THE SAMPLES

1. Draw up 15 μl of the first sample (untreated λ DNA) into the pipet tip of the micropipetter.

2. Place the end of the pipet tip beneath the running buffer, near the top of well 2 in the agarose gel. Carefully dispense the sample, which will settle to the bottom of the well. Be careful not to draw any of the liquid back into the pipet tip; i.e., do not release the button on the micropipetter until the pipet tip is out of the liquid.

3. Using fresh pipet tips, dispense 15 μl of samples 2-4 in wells 3-5. After the samples have been dispensed, proceed immediately to electrophoresis.

ELECTROPHORESIS

1. With the power supply unplugged and turned off, place the cover on the electrophoresis chamber. Check that the positive and negative leads from the chamber are connected to like terminals of the power supply.

2. Plug in the power supply and then turn it on.* Set the voltage at 60 volts.

3. After 90 min of electrophoresis, observe the blue tracking dye, *being careful not to touch the electrophoresis chamber*. To view the dye, look at the gel from the side; examining the gel from above is difficult because of condensation on the chamber cover. When the leading edge of the dye has migrated about 60 mm, switch off and then disconnect the power supply. Remove the cover of the electrophoresis chamber and proceed to staining.

STAINING

1. Wearing disposable gloves, remove the gel from the electrophoresis chamber and place it in a staining dish.

2. Add the Stains-all solution,† cover the dish, and then carefully wrap it in aluminum foil. Allow the gel to remain in the stain overnight, or about 16 hr.

DESTAINING

1. Wearing disposable gloves, pour off the stain and rinse the gel in the staining dish with several changes of cool tap water. The gel can be examined immediately but should be rinsed longer for better resolution of the bands. During the rinsing, the gel must be protected from direct light.

2. Wearing disposable gloves, transfer the gel to a sealable, clear plastic bag. Wrap the sealed bag in aluminum foil and then label and refrigerate it until the

---

*CAUTION: *During electrophoresis, do not touch the electrophoresis chamber or any of the electrical connections.*
†CAUTION: *The Stains-all solution contains formamide, which is toxic.*

electropherogram can be examined. The gel should be exposed to direct light only when it is being examined.

REFERENCES

Lewin, B. 1987. *Genes*, 3rd ed., pp. 73–83, 336–342. John Wiley, New York.

Osterman, L. A. 1984. *Methods of Protein and Nucleic Acid Research*, Vol. 1, pp. 102–151. Springer-Verlag, Berlin.

Rickwood, D. and Hames, B. D., eds. 1982. *Gel Electrophoresis of Nucleic Acids: A Practical Approach.* IRL Press, Oxford, England.

Sanger, F., Coulson, A. R., Hong, G. F., Hill, D. F., and Petersen, G. B. 1982. Nucleotide sequence of bacteriophage λ DNA. *J. Mol. Biol. 162*:729–773.

NAME _____ SECTION _____ DATE _____

LAB PARTNERS _____

EXERCISES AND QUESTIONS

1. Examine the lane with λ DNA that was treated with *Xho*I. How many bands

are visible? _____ Compare this lane with the lane containing the untreated
λ DNA. How do they differ?

_____

_____

_____

2. Examine the lane with the λ DNA that was treated with *Hin*dIII. How many

bands are visible? _____ On Data Sheet 17.1, make a sketch of the bands.
Show the approximate spacing of the bands and the location of the sample well.
To the left of each band, enter the length of the fragments present in that band.
Explain why the last two bands are stained more lightly than the other four.

_____

_____

_____

3. On Data Sheet 17.2, list the fragments from the *Hin*dIII digest and calculate
log kb. Measure the distance (in mm) from the bottom edge of the sample well
to the leading edge of each band and enter on the data sheet. Make your
measurements directly on the plastic-wrapped gel.

4. Examine the lane with the λ DNA that was treated with both restriction enzymes.

How many bands are visible? _____ Compare this lane with the lane
containing the fragments from the *Hin*dIII digest. How do they differ?

_____

_____

_____

5. On Data Sheet 17.1, make a sketch of the bands in the lane with the double
digest. Show the approximate spacing of the bands and the location of the
sample well.

6. Measure the distance from the bottom edge of the sample well to the leading
edge of the two new bands in the lane with the double digest. Make your
measurements directly on the plastic-wrapped gel. What is the distance migrated

by the larger fragment? _____mm What is the distance migrated by the smaller

fragment? _____mm

PROJECT 17

7. On Graph 17.1, draw the standard curve for the data on the five smallest *Hind*III fragments: 2.0 kb, 2.3 kb, 4.4 kb, 6.6 kb, 9.4 kb. Plot log kb (ordinate) versus distance migrated (abscissa). Draw the best-fit straight line (using a ruler) for fragments less than 5 kb and the best-fit curve (using a French curve template) for the remainder of the standard curve.

8. From the standard curve, determine the lengths of the fragments in the two new bands in the lane with the double digest. Enter the values next to the appropriate bands sketched on Data Sheet 17.1.

9. What size fragments would you predict if the *Xho*I cleavage site were closer to the 0-kb end?

   _____

   What size fragments would you predict if the *Xho*I cleavage site were closer to the 48.5-kb end?

   _____

10. In which *Hind*III restriction fragment is the *Xho*I site located?

    _____

    Explain.

    _____

    _____

    _____

    _____

Data Sheet 17.1

Diagrams of the bands in the lane with λ + *Hind*III (left) and in the lane with λ + *Xho*I + *Hind*III (right).

λ + *Hind*III   λ + *Xho*I + *Hind*III

Data Sheet 17.2
Lengths of the λ DNA restriction fragments from the *Hind*III digest, log kb, and distances migrated by the bands.

Fragment length (kb)	Log kb	Distance migrated (mm)

NAME ______________________ SECTION __________ DATE ____________

LAB PARTNERS ______________________

Graph 17.1

Standard curve: plot of log kb (ordinate) versus distance migrated (abscissa).

HIGHLY REPETITIVE DNA IN THE BOVINE GENOME

INTRODUCTION

The cells of eukaryotes have orders-of-magnitude more DNA than do prokaryotes or viruses. In addition to being larger, the genome of eukaryotes is also organized very differently. For example, it is only in eukaryotes that we find *highly repetitive DNA*. This class of DNA, which typically constitutes 5 to 15% of the eukaryotic genome, contains relatively short nucleotide sequences repeated hundreds of thousands of times. Though the function of highly repetitive DNA has yet to be determined, much is known about the nature of its base sequences. The organization of the highly repetitive sequences can be studied by treating the DNA with a restriction enzyme and then determining the lengths of the fragments with electrophoresis. Recall that each restriction enzyme cleaves DNA wherever the recognition sequence occurs. Accordingly, the fragment length is a measure of the distance between adjacent recognition sequences within the highly repetitive DNA. The objective of the project is to use this methodology to estimate the spacing of a recognition sequence within the highly repetitive DNA of the bovine genome.

CLASSES OF DNA IN EUKARYOTES

In general, prokaryotic genomes are much simpler than eukaryotic genomes. For example, there is little repetition of base sequences in prokaryotes, whereas there are multiple doses of various base sequences in eukaryotes. The kinds of DNA that may be present in the eukaryotic genome are often placed in three classes, depending on the extent to which base sequences are repeated.

Unique Sequence DNA.   *Unique sequence DNA* refers to base sequences that occur in only one dose in the haploid genome. Nearly all of a prokaryotic genome consists of unique sequence DNA, whereas in most eukaryotes, some 40 to 70% of the genome is so constituted. Examples of unique sequences are the genes that code for enzymes and other functional proteins.

Moderately Repetitive DNA.   The next most plentiful class of DNA in eukaryotes is *moderately repetitive DNA*, which may constitute as much as 30% of the genome. Moderately repetitive DNA contains nucleotide sequences that are repeated anywhere from several times to 10^4 times in the haploid genome. About 50% of the moderately repetitive DNA alternates with unique sequence DNA. There is reason to believe that some interspersed repetitive sequences may function in initiating DNA replication. Other moderately repetitive sequences code for gene products that are required in large quantity. Such sequences include the genes that code for ribosomal

RNAs (rRNAs), transfer RNAs (tRNAs), and histone proteins. Even in prokaryotes (*Escherichia coli*, for example), there are multiple copies of rRNA and tRNA genes.

Highly Repetitive DNA.   Highly repetitive DNA consists of relatively short nucleotide sequences repeated 10^4 to 10^7 times in tandem. This class of DNA is restricted to eukaryotes, where it generally constitutes 5 to 15% of the genome. Highly repetitive DNA is apparently not transcribed since it does not hybridize with any cellular RNAs.

The highly repetitive fraction is sometimes referred to as *satellite DNA*. This term derives from the fact that highly repetitive DNA has a buoyant density different from that of the bulk of the nuclear DNA and sediments as a separate band in a cesium chloride density gradient. The satellite band (or bands) may have a buoyant density that is higher or lower than that of the remainder of the DNA, but is usually different enough so that it can be isolated for further analysis. Molecular analyses in many species have demonstrated that satellite DNA is concentrated at the centromeric regions of the chromosomes. (In Project 19, the satellite DNA in the human chromosome complement is identified with a cytochemical method.) The presence of satellite DNA at the centromeric region has led to conjecture that highly repetitive DNA is essential to centromere function, or facilitates the pairing of homologous chromosomes at their centromeres. These are only speculations; the function of satellite DNA is still unknown.

THE GENOME OF *BOS TAURUS*

The haploid genome of domestic cattle (*Bos taurus*) contains about 3×10^9 base pairs, or more than 60,000 times the amount of DNA in phage λ. About 60% of the bovine genome is unique sequence DNA, about 30% is moderately repetitive DNA, and about 10% consists of highly repetitive sequences (Britten and Smith, 1969). Much of the highly repetitive DNA can be isolated for further study since it sediments as four satellites. Kurnit, Shafit, and Maio (1973) have demonstrated that the calf satellite DNAs are localized at the centromeric regions of the autosomes.

NATURE OF THE HIGHLY REPETITIVE SEQUENCES

As its name implies, highly repetitive DNA consists of some fundamental sequence repeated many times in tandem. When the patterns of repeats are determined for different organisms, some striking differences emerge. In arthropods, almost all of the satellite DNA consists of repeats of a short base sequence on the order of 5 to 10 base pairs (bp). These units are repeated millions of times in tandem, suggesting that satellite DNA in arthropods arose by successive replications of some *ancestral sequence*, a unit that is 5 to 10 bp long.

The satellite DNA in mammals appears to consist of repeating units that are 200 to 300 bp long. Base sequence analysis of the repeating units has revealed the existence of a recurrent shorter sequence, which is as long as that in arthropods (5–10 bp). Could the larger unit with 200 to 300 bp have evolved from some short ancestral sequence? Studies of satellite DNA in the mouse suggest that after successive replications of an ancestral sequence, a number of mutations accumulated in what were originally identical repeats. After replication of several adjacent units, a new larger repeating unit could have been formed.

An example using letters will illustrate the process. Let *a* represent an ancestral sequence of 5 to 10 bp. After several rounds of replication, the result would be *aaaaaa*. Now, assume that mutations occur; a change of a single base pair may occur within one or more of the *a* units, changing the sequence *a* to a new sequence, *b*, in one case, and changing *a* to *c* in another. The sequence might now appear as *aabaca*.

Replications of this entire sequence would generate *aabacaaabacaaabaca* . . . . Several additional rounds of mutation and replication would yield the base sequence pattern observed in mammalian satellite DNA.

SEQUENCE ANALYSIS OF MAMMALIAN DNA WITH RESTRICTION ENZYMES

Restriction enzymes are useful for studying the organization of repetitive DNA in mammals (see Project 17, p. 226, for a more complete discussion of restriction enzymes). Each restriction enzyme recognizes a short, specific base sequence in the DNA and cleaves each strand within this recognition sequence. The DNA fragments produced are termed *restriction fragments*. If a sample of mammalian DNA is treated with a restriction enzyme whose recognition sequence occurs periodically within the highly repetitive fraction, the DNA will be cleaved at every one of these sites. Furthermore, if these sites are regularly spaced within the highly repetitive DNA, all the restriction fragments produced will be of the same length.

The purpose of this project is to determine the spacing of a short repeating sequence within the highly repetitive DNA of the bovine genome. A sample of calf thymus DNA will be treated with the restriction enzyme *Eco*RI. Bovine highly repetitive DNA was first investigated in this manner by M. R. Botchan in the early 1970s. *Eco*RI is isolated from *Escherichia coli* and has the recognition sequence

5′GAATTC3′

3′CTTAAG5′

The arrows indicate the points at which the enzyme cuts each strand. Such cleavage will occur at every one of these sites within the DNA molecule. With the technique of agarose gel electrophoresis, the restriction fragments can be separated and their lengths determined.

ELECTROPHORESIS OF DNA FRAGMENTS

Agarose gel electrophoresis of restriction fragments is discussed in Project 17, pages 227–228. For separating DNA fragments between 0.5 and ~6 kilobases (kb), a 1% agarose gel is employed. A small gel is used, about 90 mm in length, so that the electrophoretic separation can be completed in less than 2 hr. During this time, the DNA fragments separate by size; the smaller a DNA fragment, the more rapidly it migrates through the agarose gel.

DETERMINING THE LENGTHS OF RESTRICTION FRAGMENTS

The size of a DNA fragment can be determined because its electrophoretic mobility is inversely proportional to the logarithm of its molecular weight or, simply, to the logarithm of its length in kilobases. For this determination, DNA fragments of *known* size are electrophoresed simultaneously with the fragments of unknown length. Following electrophoresis, the distances migrated by the unknowns are compared with the distances migrated by the DNA length standards. In this project, you will use a commercially prepared set of standards, termed a *1 Kb DNA Ladder* (Bethesda Research Laboratories).

The DNA ladder is prepared from two extrachromosomal genetic elements, or *plasmids*. One plasmid is obtained from yeast and the other from *E. coli*. This particular set of DNA length standards is so named because the series of bands in the electropherogram resembles the steps of a ladder, with the increment in fragment length in most adjacent bands being approximately 1 kb (Fig. 18.1). The bands containing the smaller fragments (up to about 6 kb) separate well during electrophoresis. Because of the gel's small size, the bands containing the larger DNA fragments are tightly clustered and are not resolvable. The best way to identify the bands of the DNA ladder is to compare their spacing on the electropherogram with the known spacing, illustrated in Figure 18.1.

From this set of length standards, you will construct a standard curve. The standard curve is obtained by plotting log kb (ordinate) versus distance migrated (abscissa) for each distinct band in the DNA ladder. With the gel used in this project, the plot will be linear for DNA fragments up to 4 to 5 kb in length and slightly curved for larger fragments. From the standard curve, the approximate length of the restriction fragments in the *Eco*RI-treated calf thymus DNA can be determined.

PREPARING THE SAMPLES

The digest is prepared by adding the *Eco*RI and reaction buffer (for appropriate ions and pH) to an aliquot of calf thymus DNA in a micro test tube. The sample is incubated at 37° C. After 30 min, the enzymatic reaction is halted by placing the tube at 65° C and by adding a "stop" solution. The "stop" solution contains a protein denaturant, sodium dodecyl sulfate, and a tracking dye, bromphenol blue. The tracking dye moves more rapidly than all but the smallest DNA fragments, migrating at about the same rate as fragments that are 0.5 kb long. Thus, the course of electrophoresis can be followed visually, and the process can be terminated before the colorless DNA fragments reach the end of the gel. The "stop" solution also contains glycerol, which is required for proper loading of the sample in the agarose gel.

Two other samples will be electrophoresed, an aliquot of *untreated* calf thymus DNA and a small volume of the DNA ladder. To each of these samples, the "stop" solution will be added to facilitate their proper loading in the wells.

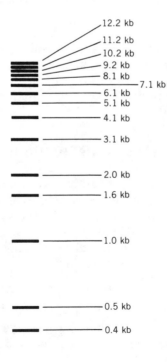

FIGURE 18.1
Electropherogram of the
1 Kb DNA Ladder
(Bethesda Research
Laboratories).

LOADING AND ELECTROPHORESING THE SAMPLES

The samples are loaded and the electrophoresis carried out, following the same procedure used in Project 17. The separation is done with a horizontal gel apparatus, in which the gel is "submerged," i.e., covered with running buffer. The samples are dispensed with an adjustable micropipetter, after which they are electrophoresed at 60 volts. Electrophoresis is halted when the tracking dye has migrated about 60 mm. The time required for such migration is 1½ to 1¾ hr. *Because of the potential for electrical shock, extreme caution must be exercised; never touch the electrophoresis chamber or any of the electrical connections while the power is turned on.*

STAINING THE DNA BANDS

The gel will be stained overnight in a 0.005% solution of Stains-all in 50% formamide. Since the stain is bleached by light, the staining dish must be wrapped in aluminum foil. The gel is then destained by rinsing with water. Destaining improves the contrast between the stained bands and the surrounding gel. The result is a series of blue bands, each containing DNA fragments of the same length.

PROCEDURES

For all procedures, you should work in teams of four. Each team will prepare and electrophorese three samples: untreated calf thymus DNA, calf thymus DNA + *Eco*RI, DNA ladder. Each team will have its own micropipetters, micro test tubes, electrophoresis chamber with agarose gel slab covered with running buffer, and power supply.

SAMPLE PREPARATION

All solutions should be kept cold until dispensed. Your instructor will demonstrate the use of the micropipetter.

1. Label three micro test tubes as shown in the following table.

Tube	Reaction buffer (10×)	Distilled water	Calf thymus DNA	*Eco*RI	DNA ladder
1	10 μl	80 μl	10 μl	—	—
2	10 μl	70 μl	10 μl	10 μl	—
3	—	—	—	—	10 μl

2. Add the solutions in the sequence given across the top of the table, from left to right. After each solution is dispensed, gently flick the tube several times to mix the contents. Add the small drop near the bottom of the tube but do not let the pipet tip touch the liquid in the bottom of the tube. (If the drop remains on the side of the tube, you will have to cap it and flick more vigorously to mix the contents.) Be sure to use a fresh pipet tip for each reagent.
3. Place tubes 1 and 2 in the micro test tube block (holder) at 37° C for 30 min. Tube 3 (DNA ladder) should be refrigerated during the incubation.
4. At the end of the incubation, add 10 μl of the "stop" solution to tubes 1 and 2 and 4 μl of the "stop" solution to tube 3. To mix the contents completely (as indicated by a *uniform* blue color), cap each tube and then swirl the contents.
5. Uncap the tubes and place them in the micro test tube block at 65° C for 5 min.

6. At the end of the incubation, remove the tubes from the block and load the samples.

LOADING THE SAMPLES

1. Draw up 15 μl of the first sample (untreated calf thymus DNA) into the pipet tip of the micropipetter.
2. Place the end of the pipet tip beneath the running buffer, near the top of well 2 in the agarose gel. Carefully dispense the sample, which will settle to the bottom of the well. Be careful not to draw any of the liquid back into the pipet tip; i.e., do not release the button on the micropipetter until the pipet tip is out of the liquid.
3. Using fresh pipet tips, dispense 15 μl of sample 2 in well 3 and 6 μl of sample 3 in well 4. After the samples have been dispensed, proceed immediately to electrophoresis.

ELECTROPHORESIS

1. With the power supply unplugged and turned off, place the cover on the electrophoresis chamber. Check that the positive and negative leads from the chamber are connected to like terminals of the power supply.
2. Plug in the power supply and then turn it on.* Set the voltage at 60 volts.
3. After 90 min of electrophoresis, observe the leading edge of the tracking dye, *being careful not to touch the electrophoresis chamber*. To view the dye, look at the gel from the side; examining the gel from above is difficult because of condensation on the chamber cover. When the leading edge of the dye has migrated about 60 mm, switch off and then disconnect the power supply. Remove the cover of the electrophoresis chamber and proceed to staining.

STAINING

1. Wearing disposable gloves, remove the gel from the electrophoresis chamber and place it in a staining dish.
2. Add the Stains-all solution,[†] cover the dish, and then carefully wrap it in aluminum foil. Allow the gel to remain in the stain overnight, or about 16 hr.

DESTAINING

1. Wearing disposable gloves, pour off the stain and rinse the gel in the staining dish with several changes of cool tap water. The gel can be examined immediately but should be rinsed longer for better resolution of the bands. During the rinsing, the gel must be protected from direct light.
2. Wearing disposable gloves, transfer the gel to a sealable, clear plastic bag. Wrap the sealed bag in aluminum foil, label, and refrigerate until the electropherogram can be examined. The gel should be exposed to direct light only when it is being examined.

*CAUTION: *During electrophoresis, do not touch the electrophoresis chamber or any of the electrical connections.*

[†]CAUTION: *The Stains-all solution contains formamide, which is toxic.*

REFERENCES

Britten, R. J. and Smith, J. 1969. A bovine genome. *Carnegie Yearb. 68*:378–386.

Kurnit, D. M., Shafit, B. R., and Maio, J. J. 1973. Multiple satellite deoxyribonucleic acids in the calf and their relation to the sex chromosomes. *J. Mol. Biol. 81*:273–284.

Lewin, B. 1987. *Genes*, 3rd ed., pp. 432–442. John Wiley, New York.

Musich, P. R., Brown, F. L., and Maio, J. J. 1977. Mammalian repetitive DNA and the subunit structure of chromatin. *Cold Spring Harbor Symp. Quant. Biol. 42*:1147–1160.

NAME _____ SECTION _____ DATE _____

LAB PARTNERS _____

EXERCISES AND QUESTIONS

1. Examine the lane with the untreated calf thymus DNA. Describe its appearance.

_____

_____

How do you explain the distribution of the stained material?

_____

_____

_____

2. Examine the lane with the *Eco*RI-treated calf thymus DNA. Locate the prominent band. Describe the appearance of the background staining in this lane.

_____

_____

3. How does the background staining in the lane with the *Eco*RI-treated calf thymus DNA differ from the staining in the lane with the untreated calf thymus DNA?

_____

_____

_____

What caused this change?

_____

_____

_____

_____

4. In the lane with the *Eco*RI-treated calf thymus DNA, measure the distance from the bottom edge of the sample well to the leading edge of the prominent band.

What is the distance? _____mm

5. Examine the lane with the DNA ladder, comparing it with Figure 18.1. Identify which bands correspond to the fragment lengths listed on Data Sheet 18.1. Measure the distance (in mm) from the bottom edge of the sample well to the leading edge of each band of the DNA ladder. Make your measurements directly on the plastic-wrapped gel. For each fragment on Data Sheet 18.1, enter the distance migrated and calculate log kb.

6. On Graph 18.1, draw the standard curve for the data on the DNA ladder. Plot log kb (ordinate) versus distance migrated (abscissa). Draw the best-fit straight line (using a ruler) for fragments less than 5 kb and the best-fit curve (using a French curve template) for the remainder of the standard curve.

7. Using the standard curve, determine the length of the DNA fragments in the

band in the *Eco*RI-treated calf thymus DNA. Length = _____kb.

8. Given your observations, what can be said about the locations of the *Eco*RI recognition sequence in the calf genome? Be as specific as possible.

_____

_____

_____

9. Various possibilities exist for the arrangement of nucleotide sequences in repetitive DNA. From your data, which of the following arrangements could

be the correct pattern in bovine highly repetitive DNA? _____ The letters *a*, *b*, and *c* represent short sequences of nucleotides.
   (i)   *baaaaaaaa...ab*
   (ii) *abababababababab*
   (iii) *abcabcabcabcabc*
   Explain.

_____

_____

_____

_____

Data Sheet 18.1

Lengths of the DNA ladder fragments, log kb, and distances migrated by the bands.

Fragment length (kb)	Log kb	Distance migrated (mm)
6.1		
5.1		
4.1		
3.1		
2.0		
1.6		
1.0		
0.5		

NAME ______________________ SECTION __________ DATE ______________

LAB PARTNERS ______________________

Graph 18.1
Standard curve: plot of log kb (ordinate) versus distance migrated (abscissa).

C-BANDING OF HUMAN CHROMOSOMES

INTRODUCTION

From the late 1960s into the 1970s, a number of laboratories developed cytochemical methods that revolutionized the field of cytogenetics. It was discovered that certain stains and pretreatments produce unique and consistent patterns of transverse bands in condensed eukaryotic chromosomes. Some of these banding procedures (G-banding and Q-banding, for example) facilitate the identification of individual chromosomes. Other banding methods are specific for certain cytological and biochemical components. The technique of C-banding differentially stains sites of constitutive heterochromatin in metaphase chromosomes. This class of heterochromatin, normally visible only at interphase and at early prophase, is concentrated at the centromeric region and is rich in highly repetitive DNA (the subject of Project 18). The objectives of this project are to induce C-bands in fixed human chromosome preparations and to identify regions of constitutive heterochromatin.

C-BANDING AND CONSTITUTIVE HETEROCHROMATIN

The phenomenon of C-banding was discovered in the course of locating certain base sequences within the mouse genome. Using the technique of *in situ hybridization*, Pardue and Gall (1970) found that satellite DNA, i.e., highly repetitive DNA, is located at the centromeric regions. Since 1970, similar findings have been reported for many species of plants and animals. These *in situ* hybridization studies involve incubation of radioactive satellite DNA (or, more commonly, radioactive RNA that is complementary to satellite DNA) with chromosome spreads that have been pretreated with alkali. The alkali denatures the chromosomal DNA, i.e., separates the two strands of the double helix, allowing the radioactive satellite DNA to hybridize with complementary sequences within the genome. The chromosome spreads are exposed to the radioactive DNA in the presence of a hot saline citrate solution, which provides the ionic conditions that favor renaturation. After the hybridization occurs, autoradiographs are made and the slides stained with Giemsa.

In the mouse cells, Pardue and Gall observed that the label was concentrated over the centromeric regions, indicating that these are the chromosomal sites of satellite DNA. In addition, the centromeric regions, which are known to contain constitutive heterochromatin, stained more deeply with Giemsa blood stain. The darker staining was of considerable interest since constitutive heterochromatin is normally visible only from interphase through early prophase. An important spin-off of the Pardue and Gall study was the observation that with alkali pretreatment of cells and subsequent incubation in a hot saline citrate solution (without labeled nucleic acids), ordinary Giemsa stain could be used to identify constitutive heterochromatin in *condensed* (metaphase) chromosomes. The procedure, which became known as C-banding because of its specificity for constitutive heterochromatin, was refined by Arrighi and Hsu (1971).

Constitutive heterochromatin is found in the chromosomes of essentially all eukaryotes. Most of it is found adjacent to the centromeres, though it sometimes occurs at telomeres (tips of the chromosomes) and at interstitial sites. In some organisms, most notably *Peromyscus* (the deer mouse), entire chromosome arms are composed of constitutive heterochromatin. While the function of constitutive heterochromatin is not understood, there is some evidence that it plays a role in the regulation of crossing over and, possibly, in chromosome pairing and segregation.

Mechanism of C-Banding. The mechanism by which C-bands are induced has been the subject of much research (see review by Comings, 1978). The differential staining appears to be related to the amount of DNA remaining in different chromosomal regions after the C-banding pretreatments. It has been demonstrated that treatment with alkali not only denatures DNA but removes about 60% of the chromosomal DNA. Most of the DNA is extracted from the non–C-band regions, suggesting that C-bands stain more deeply simply because they contain more DNA after the extraction. Why the DNA in C-banded regions is more resistant to extraction remains to be determined.

HUMAN CHROMOSOME PREPARATIONS

The source of the chromosomes for this investigation is the short-term human leukocyte culture. Human leukocytes are grown in a rich culture medium. After 24 hr in culture, the small lymphocytes commence DNA synthesis and subsequently undergo mitosis. Peak mitotic activity occurs between 60 and 72 hr in culture. To accumulate a large number of metaphases, *colchicine* is added for the final 4 hr of culture. Colchicine interferes with spindle formation so that dividing cells do not proceed through mitosis but, instead, are arrested at a stage termed *c-metaphase* (*c* is for colchicine).

Following the incubation with colchicine, the cells are collected from the culture by centrifugation. The basic steps in the harvest are treatment with a hypotonic solution, fixation of the cells, and air drying of the fixed cell suspension on slides. The hypotonic solution (0.075 M KCl) swells the cells and disperses the chromosomes. Use of a rapidly penetrating and volatile fixative (1:3 acetic:methanol) also facilitates spreading of the chromosomes. The cells are placed on a microscope slide and rapidly air dried, leaving all the chromosomes in one focal plane. You will receive fixed slide preparations that are ready for the C-banding procedure.

THE HUMAN CHROMOSOME COMPLEMENT

The normal human chromosome complement consists of 22 pairs of autosomes and a pair of sex chromosomes, XX in females and XY in males. The *karyotypes* of a normal human female and a normal human male are shown in Figure 19.1. A karyotype is an ordered arrangement of the chromosomes prepared from an enlarged photograph. The numbering and grouping of the chromosomes are determined using standard rules agreed upon by cytogeneticists. The chromosome pairs are serially numbered by decreasing total length and placed in seven groups (A–G) by centromere location or *arm ratio*, defined as length of long arm/length of short arm. On the basis of size and arm ratio, every chromosome can unequivocally be placed into one of the seven groups. In conventionally stained preparations, only a few chromosomes can be identified individually. In Figure 19.1, these chromosomes have a number directly beneath them; the other chromosomes are designated by group only.

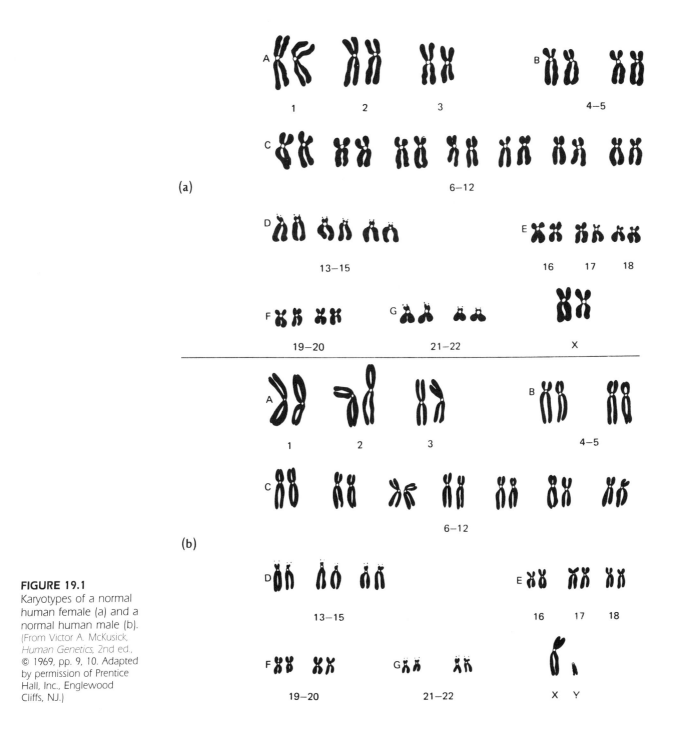

FIGURE 19.1
Karyotypes of a normal human female (a) and a normal human male (b). *(From Victor A. McKusick, Human Genetics, 2nd ed., © 1969, pp. 9, 10. Adapted by permission of Prentice Hall, Inc., Englewood Cliffs, NJ.)*

The features of the seven groups are summarized in Table 19.1. The largest human chromosomes are in the A group, about 8 μm in an average metaphase spread. Each member of the A group can always be identified. Chromosomes A1 and A3 are metacentric, but A1 is appreciably longer; A2 is the largest submetacentric (see discussion of chromosome morphology in Project 4, p. 30). The B-group chromosomes are large submetacentrics. They are only slightly longer than the largest members of the C group but are easily identified by their greater arm ratio. The C-group chromosomes are medium submetacentrics and show a gradation in size. Members of the D group are medium acrocentrics with a secondary constriction and satellite in the short arm. The chromosomes in the E group are all short submetacentrics, yet each can be identified by its centromere location; E16 is nearly metacentric, while E17 and E18 are progressively more submetacentric. The F-group chromosomes are small metacentrics, and the G-group chromosomes, which are the smallest in the complement (about 2 μm), are acrocentrics.

TABLE 19.1
General characteristics of the chromosome groups in the human karyotype

Group	Chromosome number	Description
A	1,3	Large metacentrics
	2	Large submetacentric
B	4,5	Large submetacentrics
C	6–12,X	Medium submetacentrics
D	13–15	Medium acrocentrics with satellites
E	16–18	Small submetacentrics
F	19,20	Small metacentrics
G	21,22	Small acrocentrics with satellites
	Y	Small acrocentric; lacks satellite

In conventionally stained preparations, i.e., Giemsa staining without any pretreatment, the X chromosome is difficult to distinguish from the larger members of the C group and, accordingly, is grouped with them. The Y chromosome is sometimes placed in the G group, though the Y can often be identified by the orientation of its long arms, which tend to be less divergent than the long arms of chromosomes 21 and 22. In addition, the Y lacks the secondary constriction and satellite present in the autosomes of the G group. Although shown in the stylized karyotypes of Figure 19.1, the secondary constriction and satellite in each acrocentric autosome may not always be apparent.

For positive identification of every human chromosome, cytogeneticists rely on chromosome banding methods, such as G-banding and Q-banding. These staining techniques produce a distinctive pattern of transverse dark and light bands along the length of each chromosome. With the technique of C-banding, constitutive heterochromatin is differentially stained and several human chromosomes can be identified from the size of their C-bands.

Eye Karyotype Drawing. The chromosome analysis to be carried out on conventionally stained preparations is an *eye karyotype drawing*, which is depicted in Figure 19.2. Every chromosome in the metaphase spread is represented by a line and subsequently identified by group or number. The eye karyotype method is used in clinical evaluations and is an excellent way to become familiar firsthand with the human karyotype.

C-BANDS IN HUMAN CHROMOSOMES

The C-banded human karyotype was first described by Arrighi and Hsu in 1971. Every chromosome in the human complement is deeply stained at the centromeric region, and the Y chromosome is deeply stained in the distal region of the long arm. Several human chromosomes can be identified unequivocally in C-banded preparations. Chromosomes 9 and 16 have especially large C-bands, and, as noted, the Y chromosome has much of its long arm deeply stained. In human populations, there is observed a polymorphism in the size of C-bands, particularly in chromosomes 1, 9, and 16. Variation in the size of C-bands has also been observed in chromosomes 12, 17, 19, and 21 (see discussion and photomicrographs in Therman, 1986, pp. 211–213).

As is observed in other species, the C-bands in human chromosomes contain highly repetitive DNA. The highly repetitive DNA in the human genome sediments as four satellites, each with a characteristic buoyant density and base sequence. An *in situ* hybridization study by Gosden *et al.* (1975) has determined that the C-bands in human chromosomes exhibit great variation in the proportions of the satellite DNAs they contain. For example, the C-bands in chromosome 9 and the long arm of the Y contain all four satellite DNAs, whereas the C-band in chromosome 16 contains only satellite II.

THE C-BANDING PROCEDURE

The C-banding protocol selected for this project is a modification of a procedure developed by Sumner (1972). You will be processing fixed, air-dried slide preparations derived from human leukocyte cultures. The slides will already have been treated with a dilute solution of hydrochloric acid for 30 min. The HCl removes proteinaceous debris that might otherwise detract from the C-banding. You will begin the procedure with the alkali treatment; the slide preparations are immersed briefly in a hot, saturated solution of barium hydroxide, $Ba(OH)_2$. After a thorough rinsing in distilled water, the slides are incubated in a hot, concentrated ($2\times$) solution of *standard saline citrate*, abbreviated $2\times SSC$. The incubation is carried out in a *moist chamber*, as illustrated in Figure 19.3. The slides are then stained with Giemsa, after which a coverslip is mounted.

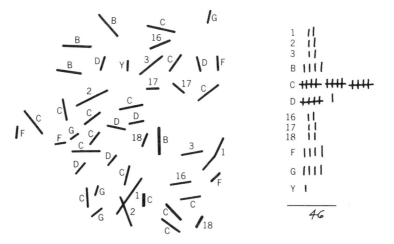

FIGURE 19.2

An eye karyotype drawing. (From J. H. Priest, 1977, *Medical Cytogenetics and Cell Culture*, 2nd ed., p. 188, Lea & Febiger, Philadelphia.)

PROCEDURES

For all staining procedures, you should work in pairs. Each team will process two slides, sharing a moist chamber. While the slides are incubating in the 2×SSC, each member of the team can examine a prepared, conventionally stained slide of human chromosomes.

1. Label the unstained slide preparations and prepare a moist chamber for step 4. Place four rubber washers in a square Petri dish, as shown in Figure 19.3. Add a small quantity of 2×SSC to cover the bottom of the dish.

2. Wearing disposable gloves, remove the scum from the surface of the hot (50° C), saturated solution of $Ba(OH)_2$ by drawing a piece of filter paper across the surface of the solution. Dispose of the filter paper properly.

3. Using forceps, immerse the slides, one at a time, in the Coplin jar of hat $Ba(OH)_2$ for 15 sec. *Immediately* rinse each slide thoroughly in several changes of distilled water in a Coplin jar. Finally, dip the slide briefly in a Coplin jar of 2×SSC.

4. Place the slides on the washers in the moist chamber. Using a Pasteur pipet, flood the surface of each slide with 2×SSC. Place a large coverslip on each slide, cover the Petri dish, and then place it in an oven at 60° C for 1 hr. While the slides are incubating in the 2×SSC, examine a conventionally stained slide of human chromosomes and prepare an eye karyotype drawing, as described in Observations and Questions, steps 1–5.

5. After the 1-hr incubation, remove the coverslips and then rinse the slides in several changes of distilled water in a Coplin jar.

6. Stain the slides for 20 min in a freshly prepared 4% Giemsa solution in 0.01 M phosphate buffer, pH 6.8. Convenient volumes for a Coplin jar are 36 ml phosphate buffer and 1.5 ml Giemsa stain.*

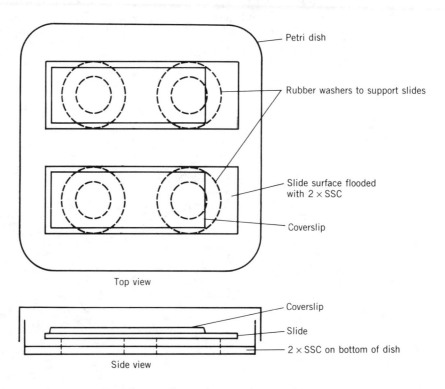

FIGURE 19.3
Moist chamber for the incubation in 2× standard saline citrate (2×SSC).

*CAUTION: *Do not pipet by mouth.*

7. Rinse each slide under running tap water for several seconds and allow to air-dry in a near-vertical position. When each slide is completely dry, dip briefly in xylene (in a fume hood) and mount a coverslip with two drops of Permount. Lower the coverslip slowly to minimize air bubbles and then gently press out the excess liquid with a paper towel.

REFERENCES

Arrighi, F. E. and Hsu, T. C. 1971. Localization of heterochromatin in human chromosomes. *Cytogenetics 10*:81–86.

Comings, D. E. 1978. Mechanisms of chromosome banding and implications for chromosome structure. *Ann. Rev. Genet. 12*:25–46.

Gosden, J. R., Mitchell, A. R., Buckland, R. A., Clayton, R. P., and Evans, H. J. 1975. The location of four human satellite DNAs on human chromosomes. *Exp. Cell Res. 92*:148–158.

Pardue, M. L. and Gall, J. G. 1970. Chromosomal localization of mouse satellite DNA. *Science 168*:1356–1358.

Priest, J. H. 1977. *Medical Cytogenetics and Cell Culture*, 2nd ed. Lea & Febiger, Philadelphia.

Sumner, A. T. 1972. A simple technique for demonstrating centromeric heterochromatin. *Exp. Cell Res. 75*:304–306.

Therman, E. 1986. *Human Chromosomes*, 2nd ed., pp. 13–24, 32–57, 211–213. Springer-Verlag, New York.

OBSERVATIONS AND QUESTIONS

1. On the conventionally stained slide, locate a well-spread cell at c-metaphase. Using the oil-immersion objective, count the chromosomes. A cross-hair reticle in the ocular facilitates the counting, since there are four distinct regions with an easily scored number of chromosomes in each quadrant. The number of chromosomes is considered verified when the same total is obtained in two successive counts.

2. For a cell with 46 chromosomes, prepare an eye karyotype drawing on Data Sheet 19.1. Use Figure 19.2 as a guide. Make a line representation of each chromosome. Retain the relative positions of the chromosomes in the spread and their relative lengths. Simple lines are adequate since chromosomes will subsequently be identified from microscopic examination.

3. Next to the sketch, make a list of the chromosome numbers and groups, as shown in Figure 19.2. Identify the members of the A group in the microscope and label each by number on your sketch. As each is located, enter a tally mark on your list.

4. Following the same procedure as with the A group, locate (microscopically and on your sketch) *in sequence* the group B, G, F, D, E, and C chromosomes. Try to identify pairs 16, 17, 18, and the Y (in a male).

5. Total the number of chromosomes.

6. Suggest a much simpler method than making an eye karyotype drawing to determine whether a slide preparation is from a male or a female.

_____

_____

_____

7. On the C-banded slide, locate a cell in which the chromosomes are adequately spread and there is good C-banding, i.e., deeply stained centromeric regions and lightly stained chromosome arms.

8. Identify and draw (on p. 267) chromosome pairs 1, 9, 16, and the Y (in males).

   Which of these autosomes (1, 9, or 16) has the *largest* C-band? _____ Which

   of these autosomes has the *smallest* C-band? _____ In which of these autosomes are there unequal amounts of centromeric heterochromatin in the two arms?

   _____

9. Is there a difference in the size of the C-band in the two homologues of pairs 1,

   9, and/or 16? _____ How might such variation occur?

   _____

   _____

Data Sheet 19.1
Eye karyotype drawing.

DNA REPLICATION

INTRODUCTION

The *basic* mechanism of DNA replication was first suggested by J. D. Watson and F. H. C. Crick as part of their work on the structure of DNA. In the replication process, the two strands of the DNA double helix (or duplex) separate, and each original strand serves as a template for the synthesis of a complementary strand. The term *semiconservative* is used to describe this mode of replication, since one strand, i.e., *half* of the parental duplex, remains intact in each daughter duplex. The first evidence indicating that the DNA in eukaryotic chromosomes replicates semiconservatively was provided by the autoradiographic experiments of J. H. Taylor and coworkers in 1957. DNA replication in eukaryotic chromosomes can also be followed cytochemically using a special dye, *Hoechst 33258*. The objective of this project is to demonstrate semiconservative replication with Hoechst–Giemsa staining.

BrdU–HOECHST METHOD

Hoechst 33258 is a *fluorescent* chromosome stain; the dye molecules bind to the DNA, and the chromosomes fluoresce brightly. In 1973, S. A. Latt determined that the fluorescence is reduced if the base analogue *5-bromodeoxyuridine (BrdU)* is incorporated in the DNA. BrdU resembles thymidine except that at the 5 position of the pyrimidine ring there is a bromine atom instead of a methyl group. Latt grew human lymphocytes in medium containing BrdU, which is incorporated in the chromosomal DNA in lieu of thymidine. He then stained the cell preparation with Hoechst 33258. In cells that had undergone one replication in the BrdU medium, both sister chromatids fluoresced brightly, though not as brightly as chromosomes without *any* BrdU. After two replications in the BrdU medium, one sister chromatid fluoresced brightly and the other fluoresced weakly. Other experiments revealed that the weakly fluorescing chromatid contained about twice as much BrdU as its brightly fluorescing sister chromatid. Thus, the brightly fluorescing chromatid has BrdU incorporated in only one strand of its DNA, while the weakly fluorescing sister chromatid has BrdU incorporated in both strands. As shown in Figure 20.1a, these observations are precisely what would be predicted if an unreplicated chromosome contains one DNA duplex, and replication proceeds semiconservatively. An additional observation at the second metaphase is the presence of "harlequin" chromosomes (Fig. 20.1b). The alternating pattern of bright and dull regions along each chromatid results from *sister chromatid exchanges (SCEs)*, i.e., the breakage and reunion of homologous segments of sister chromatids.

269

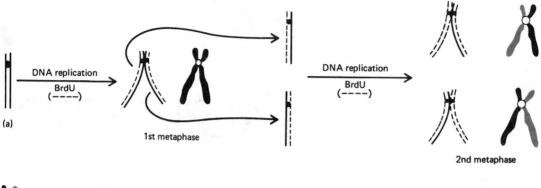

FIGURE 20.1

(a) Incorporation of 5-bromodeoxyuridine (BrdU) in DNA and the subsequent appearance of the chromosomes after staining with the fluorescent dye Hoechst 33258. Chromatids in which one DNA strand is substituted with BrdU fluoresce brightly (deeply shaded), whereas chromatids in which both DNA strands are substituted with BrdU fluoresce weakly (lightly shaded). (b) Formation of a chromosome with two sister chromatid exchanges (at arrows) and its appearance at the second metaphase.

HOECHST–GIEMSA STAINING

The BrdU–Hoechst technique, as done by Latt, has a technical problem in that the fluorescence fades rather quickly. This difficulty was obviated by Wolff and Perry (1974), who found that if the Hoechst-stained slide is exposed to light for 24 hr, the BrdU-substituted chromatids at the second metaphase can be differentially stained with ordinary Giemsa blood stain. With this procedure, known as the *fluorescence-plus-Giemsa (FPG)* method, the chromatid with one BrdU-substituted strand of DNA stains deeply, while the chromatid with two BrdU-substituted strands stains lightly. Apparently, the exposure to light interferes with the binding of the Giemsa dye to the doubly substituted, Hoechst-stained chromatid. Not only is the Giemsa-stained preparation resistant to fading, but it is also excellent for detecting sister chromatid exchanges. There is currently great interest in SCEs, for they are an indication of genetic damage. Accordingly, Hoechst–Giemsa staining is used for screening potential mutagens and carcinogens (Wolff, 1982).

The protocol selected for the present study was developed by Goto *et al.* (1975). Human leukocytes are cultured for two to three cell generations in medium containing BrdU. The principles underlying the culturing of human blood cells are discussed in Project 19, page 256. You will be provided with unstained slide preparations from such harvested cultures.

The first step in the FPG staining procedure is brief treatment of the unstained slide preparations with a relatively high concentration of Hoechst 33258. By using a high concentration of the dye, Goto and coworkers found that the time required for light exposure is dramatically reduced. After a 1-hr exposure to an intense fluorescent light source, the slide preparations are stained with Giemsa blood stain. The last step is the mounting of a coverslip with Permount.

PROCEDURES

HOECHST–GIEMSA STAINING

Always use forceps to place slides in or to remove slides from Coplin jars.

1. Dip a labeled slide briefly in distilled water and then place in an aqueous solution of Hoechst 33258, 200 μg/ml, for 10 min.

2. With forceps, transfer the slides to an empty Coplin jar and then rinse the slides under running tap water for 2 min.
3. Replace with distilled water, pour it off, and then prop the slides in a near-vertical position until dry or nearly so.
4. With a Pasteur pipet, flood the surface of the slides with McIlvaine's buffer. Place a large coverslip on each slide.
5. Place the slides approximately 3 cm from two fluorescent bulbs. Expose the slides to the light for 1 hr, checking occasionally that the coverslips are in place.
6. Remove the coverslips and then rinse each slide under running distilled water for several seconds.
7. Stain the slides for 6 min in a freshly prepared 4% Giemsa solution in 0.01 M phosphate buffer, pH 6.8. Convenient volumes for a Coplin jar are 36 ml phosphate buffer and 1.5 ml Giemsa stain.*
8. Rinse each slide under running tap water for 5 sec and allow to air-dry in a near-vertical position. When each slide is completely dry, dip briefly in xylene (in a fume hood) and mount a coverslip with two drops of Permount. Lower the coverslip slowly to minimize air bubbles and then gently press out the excess liquid with a paper towel.

REFERENCES

Goto, K., Akematsu, T., Shimazu, H., and Sugiyama, T. 1975. Simple differential Giemsa staining of sister chromatids after treatment with photosensitive dyes and exposure to light and the mechanism of staining. *Chromosoma 53*:223-230.

Latt, S. A. 1973. Microfluorometric detection of deoxyribonucleic acid replication in human metaphase chromosomes. *Proc. Nat. Acad. Sci. 70*:3395–3399.

Wolff, S., ed. 1982. *Sister Chromatid Exchange*. John Wiley, New York.

Wolff, S. and Perry, P. 1974. Differential Giemsa staining of sister chromatids and the study of sister chromatid exchanges without autoradiography. *Chromosoma 48*:341–353.

*CAUTION: *Do not pipet by mouth.*

OBSERVATIONS AND QUESTIONS

1. Scan the slide for a cell at the first metaphase. How do you know that the cell has undergone only one division in the BrdU medium?

   _____

   _____

   _____

2. Scan the slide for a cell at the second metaphase. On page 275, draw several chromosomes in this spread, shading the chromatids to show their staining intensity.

3. From the cell in question 2, draw the chromosome that shows the greatest number of SCEs. How many reciprocal breaks must have occurred to produce

   the observed configuration? _____

4. Scan the slide for a cell in which some chromosomes have differentially stained chromatids and some have both chromatids lightly stained. How many

   replications in BrdU did this cell undergo? _____ Explain your answer using diagrams similar to those in Figure 20.1a.

5. Using diagrams, explain the origin of a chromosome with the following appearance:

PHOTOMICROGRAPHY

INTRODUCTION

Photomicrography is a valuable adjunct to microscopy, for it provides the cell biologist with an accurate, permanent, visual record of microscopic observations. Prior to the invention of photography, the only means of recording such images was by sketching them. In 1850, even though photography was a primitive science, acceptable photomicrographs were being made on wet plates. With the advent of dry emulsions, specially designed photomicrographic cameras, and improved microscope optics, taking quality photomicrographs became a routine matter. The visual record takes two forms, the *print* for published works and the *transparency* for projection to an audience. Prints and transparencies can be made with "instant" films or with traditional wet-process methods, which involve chemical processing in a darkroom. Included in this project are protocols for making black-and-white prints, either with the wet-process method or with instant film. Also included is a protocol for making black-and-white transparencies with instant 35mm film. The objective of this project is to learn techniques that will enable you to photograph a variety of preparations you have made in the cell biology laboratory.

PRINCIPLES OF PHOTOMICROGRAPHY

It is possible to record microscopic observations on film because, in addition to the virtual image perceived by the eye, there is a real image projected through the ocular. When a camera is placed above the ocular, generally the photographic ocular of a *trinocular microscope*, the image can be captured on film. Before any photomicrography is done, the microscope must be adjusted for Köhler illumination (Project 1, p. 7), with the condenser iris set for optimum contrast. Regardless of how sophisticated the camera, the quality of the photomicrograph will be less than ideal if the microscope is improperly adjusted.

The basic steps for taking photomicrographs are the same as in pictorial photography. First, the picture is composed; the cells or structures to be photographed are positioned in the field so that they are included in the film frame. Next, the image is brought into sharp focus at the film plane. Finally, the correct exposure is determined. These steps are described for two popular film formats, 35mm film and Polaroid pack film.

Composition and Magnification. The picture is composed with the aid of a *focusing eyepiece* (Fig. 21.1). This special ocular contains a reticle with one or more frames, indicating the region of the microscope field that will be included on the film frame with different size films. The 35mm frame is 24mm × 36 mm and the Polaroid print frame is 8.3 cm × 10.8 cm (3¼ in. × 4¼ in.). When the field you would like to photograph is larger than the reticle frame, the total magnification should be reduced. By turning the focusing eyepiece, you can select the regions of the field that will be

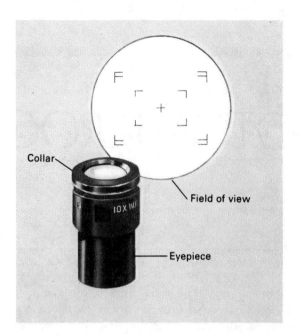

FIGURE 21.1
Focusing eyepiece and its focusing/framing reticle. (Courtesy of Cambridge Instruments.)

Collar

Field of view

Eyepiece

included in the frame. When good composition is achieved, the camera is turned so that it is positioned at the same angle as the reticle frame.

The magnification at the film plane is determined by the particular objective and photographic ocular being used. If prints are to be made, it is desirable to photograph the calibrations of a stage micrometer slide. This negative will later be printed.

Focusing.   The focusing eyepiece enables you to focus the microscope image at the film plane of the camera. First, the collar on the focusing eyepiece is turned until the cross hairs on the reticle (Fig. 21.1) are in sharp focus. Then, the image is brought into focus with the fine adjustment knob. If you have difficulty focusing on the cross hairs, it may be helpful to gaze at a distant object before focusing.

Determining the Correct Exposure.   In all photography, the *exposure (E)* is determined by *light intensity (I)* and *exposure time (T)*, the relationship being $E = IT$. Thus, the exposure remains the same if, for example, the light intensity is halved and the exposure time simultaneously doubled. In photomicrography, the intensity of the light that reaches the film is affected by many factors. These include the transformer setting, which determines the voltage reaching the illuminator bulb, any filters in the light path (see next section), the magnifications of the objective and ocular, and, of course, the nature of the specimen. Generally, one controls light intensity by changing the transformer setting or the filter in the light path.

The exposure time is controlled by the shutter setting. As in pictorial photography, an *exposure meter* is a great aid in determining the correct exposure time. The meter must be adjusted for the film's speed rating, given as an *ISO* number (same as ASA). The ISO number is an arithmetic measure of the sensitivity of the film to light; the higher the ISO designation, the "faster" or more light-sensitive is the film. For example, a film rated at ISO 100 requires one-half the exposure of a film rated at ISO 50. Some exposure meters must also be adjusted for the film format (35mm or 8.3 × 10.8 cm) and the magnification of the photographic ocular. The meter reading, hereafter termed the *indicated exposure time*, is only an estimate of the best exposure time. If the indicated exposure time is greater than the slowest available shutter speed on your photomicrographic assembly, obviously you will have to turn up the transformer setting.

To obtain the best exposure time, it is necessary, at first, to run a series of exposures with successively doubled exposure times. The exposure series should bracket the indicated exposure time. Assume, for example, that the indicated exposure time is 1/15 sec. Then, a series of three 35mm frames can be exposed at 1/30, 1/15, and 1/8 sec, and you can be fairly certain of having one properly exposed frame. For Polaroid instant prints, where each frame is more costly, an exposure series can be made on *one* print if your camera back has a *dark slide* close to the film plane. The multiple exposures on the single print can then be examined to decide on the best exposure time for the final print.

After you have made an exposure series with 35mm film, you will have to determine which frame has the best exposure. Selecting which of several transparencies has the best exposure is an easy matter since it is similar to selecting a good print. Selecting the best negative in an exposure series is more difficult for the beginner because the negative has the *reverse* light–dark features of the subject. You should check to see which frame has the greatest detail in *light* areas on the negative. If the negative is underexposed, there will be very little detail in such areas. If the negative is greatly overexposed, you will not be able to see newsprint through the darkest areas of the negative. Since negative films have considerable exposure latitude, it is likely that more than one frame will be printable. However, in choosing between a slightly overexposed and a slightly underexposed negative, it is better to print the former.

Filters. To enhance contrast in black-and-white photomicrography, we use filters with colors that are complementary to those of the stained specimen. For the red and purple hues in stained cells and tissues, a green filter, such as a Kodak Wratten Filter No. 58, or a yellow-green filter, such as a No. 11, should be used. The green filtration also minimizes the effects of lens aberrations since objectives are optically corrected in this spectral region.

If the light intensity is very high, as would be the case at low magnification, the correct exposure time might be shorter than the fastest available shutter speed. Light intensity can be reduced by lowering the transformer setting or by inserting a *neutral density filter*. Commonly used neutral density filters have transmittances of 50%, 25%, and 12.5%.

Checklist. With so many factors affecting the quality of each photomicrograph, it is essential to keep a record of important variables. A checklist for your personal record is provided on Data Sheet 21.1. If you have kept a careful record, you will know the best exposure time for each combination of film, stain, and filter. Such data will be helpful for taking photomicrographs in the future.

BLACK-AND-WHITE, WET-PROCESS METHODS

35mm Film. An excellent 35mm film for routine black-and-white photomicrography is Kodak *Technical Pan Film 2415*. It has extremely high resolving power and also has extremely fine grain, permitting the negative to be printed with considerable enlargement. With the processing protocol given in this project, Technical Pan has a rating of ISO 50 and yields negatives of moderate contrast. Like most films, Technical Pan is *panchromatic*, i.e., sensitive to all wavelengths in the visible spectrum, and must be handled in complete darkness.

Film Processing. All films consist of two basic layers. There is a light-sensitive *emulsion*, which is a layer of gelatin containing crystals of silver halides, and a *base*, which is a polyester or cellulose acetate support surface. The silver halides are very light-sensitive, and even a small amount of light reduces some of the silver ions to metallic silver in small regions of the crystals. The image formed is not visible to the eye and is termed a *latent image*. To produce a *visible image*, the film must be *developed*. The developer chemically reduces more of the silver ions in the exposed crystal, producing a *grain* of metallic silver. Development is arrested with a *stop bath*, a weak solution of acetic acid that neutralizes the alkaline developer. Following development, the film is *fixed*, or made permanent, with *fixer*, which contains sodium hyposulfite and a chemical that hardens the gelatin coat. The sodium hyposulfite makes the image permanent by converting the unexposed silver halide crystals to soluble compounds, which can then be rinsed away. The finished negative has the reverse light–dark features of the subject, for there are now dark areas with silver grains corresponding to light areas on the subject and clear areas with no silver grains corresponding to dark areas on the subject. In the second phase of wet-process photography, *printing*, the reversed light–dark image on the film is itself reversed on the *print*.

Print Paper.   Print paper is heavy white paper with an emulsion of silver halides on one surface. Unlike most films, print paper can be handled using a *safelight*, which is a weak light source covered with a colored filter that transmits light in a spectral region where the emulsion is not very sensitive. The print paper used in this project is Kodak *Polycontrast III RC Paper*. The term *polycontrast* indicates that different degrees of print contrast are obtained with different Kodak *Polycontrast II* (PC) filters used during printing. The PC filters are numbered in half steps from 0 through 5; the higher the number, the greater the contrast of the print. For this project, you can start with a PC 2 filter. The RC paper is available in two finishes, *glossy* (designated F) and *matte* (designated N). The glossy finish is used for research photomicrographs since journals require a glossy finish for publication.

Printing.   In the printing process, light is allowed to pass through the negative onto the print paper. Generally, we want to enlarge the image on the negative, a process requiring a *photographic enlarger* (Fig. 21.2). The main components of an enlarger are the lamp, filter drawer, condenser, negative carrier, and focusable lens system. When the negative is placed in the negative carrier and the enlarger lamp turned on, the image is projected onto the print paper below. The print paper is held flat in an *enlarger easel* placed on the baseboard. As light strikes the paper's emulsion, a latent image is formed. Just as with film, a developer is required to convert the latent image to a visible image, and fixer is required to remove the unexposed silver halide crystals, making the print permanent. In printing, it is necessary, once again, to compose the picture, focus the image, and determine the correct exposure.

Composition and Magnification.   When composing the print, you will have to obtain an appropriate degree of enlargement. The degree of enlargement is controlled with the height control knob (Fig. 21.2). The height may have to be adjusted several times until all the areas of interest appear in a projection of reasonable size. The areas of interest are framed using the enlarger easel; the easel can be moved around on the baseboard, and the movable borders of the easel can be adjusted. To obtain the exact magnification, the negative of the stage micrometer scale should be projected and later printed. Magnification is computed as follows:

$$\text{magnification} = \frac{\text{distance between scale divisions on the print}}{\text{distance between scale divisions on the stage micrometer}}$$

Focusing.   The focusing is done by adjusting the length of the enlarger bellows, which changes the distance between the negative and the lens. You will have to adjust the focus whenever the magnification is changed. The negative is in focus when the grains of the negative's emulsion appear in sharp focus on the print paper. For viewing the grains, you will need an *enlarger focusing scope* to magnify the image.

Determining the Correct Exposure.   Exposure, you will recall, is the product of light intensity and exposure time. In printing, light intensity is controlled by the enlarger lens opening, expressed as the ratio of lens focal length/diameter of the lens

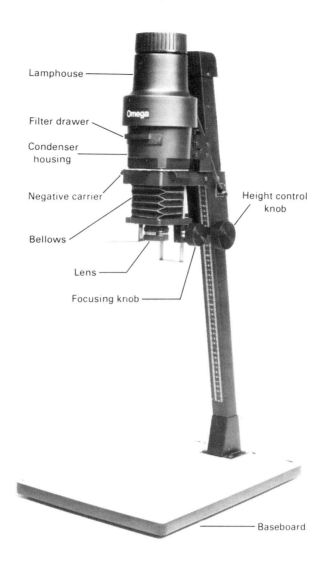

FIGURE 21.2
Photographic enlarger.
(Courtesy of RT/Omega
Industries, Inc.)

Lamphouse

Filter drawer

Condenser
housing

Negative carrier

Bellows

Lens

Focusing knob

Height control
knob

Baseboard

aperture. This ratio is commonly referred to as the *f-stop*. Since the focal length of the lens is constant, a higher *f*-stop indicates a smaller lens aperture. In a typical series of *f*-stops (*f*/4, *f*/5.6, *f*/8, *f*/11, *f*/16, *f*/22), each higher *f*-stop lets through half as much light as does the preceding *f*-stop. Since $E = IT$, the same exposure would be obtained with 5 sec at *f*/8, with 10 sec at *f*/11, or with 20 sec at *f*/16.

There is a relatively simple method for determining the correct exposure for a particular negative. The procedure relies on a *projection print scale*, which is a sheet of celluloid divided into sectors. Each sector is in a different shade of gray and has a number (of seconds) printed on it. With the negative in the negative carrier, the appropriate PC filter in the filter drawer, the projection print scale overlying a fresh piece of print paper, and the enlarger lens aperture set at *f*/8, a 1-min exposure is made. The print is developed and the sector with the best exposure selected. The number in the sector with the best exposure is the approximate number of seconds (at *f*/8) required to obtain a print with that exposure. If the indicated exposure time is too short to gauge accurately, for example, 3 sec at *f*/8, you might want to expose the first print for 12 sec at *f*/16. A convenient range for exposure time is 10–15 sec. Should you want to print the same negative at a different magnification or at a different contrast, you will have to make a new exposure series with the projection print scale.

INSTANT-PROCESS METHODS

Black-and-White Prints.   A popular instant print film for photomicrography is Polaroid[1] *Type 667*. The film is very fast (ISO 3000) and, hence, rather grainy. With this pack film, a print is produced in about 1 min. The chemical processing begins automatically as the print is removed from the special camera back. Since there is no negative, print magnification is limited by the available microscope lenses, and print contrast can be controlled only with the condenser iris of the microscope and with colored filters.

Transparencies.   Black-and-white transparencies can also be processed quickly without a darkroom, using Polaroid *PolaPan*[2] 35mm film. This film, which can be used in any 35mm camera, is a *reversal film*, meaning that it yields positive transparencies directly. PolaPan is a continuous-tone, panchromatic, black-and-white film with high resolution, fine grain, and medium speed (ISO 80).

After the film has been exposed and rewound, the cartridge with its film leader is placed in the *Polaroid 35mm AutoProcessor*, which is a "black box" with gears, rollers, lever, and crank. Also placed in the processor is a *processing pack*, which contains the developing chemicals. Turning the crank brings the film and developing chemicals into contact. Development is allowed to proceed for 60 sec, after which the film is rewound and removed from the processor. The film is then completely dry and can immediately be cut, mounted, and projected.

PROCEDURES

PHOTOGRAPHING CELL PREPARATIONS WITH TECHNICAL PAN FILM

1. Load the film in the 35mm camera back and advance the film three frames. After the camera is loaded and mounted, pull out the dark slide.

2. Place the microscope slide on the stage and adjust the microscope for Köhler illumination (see Project 1, p. 7). After you have located the field you want to photograph, adjust the condenser iris for optimum contrast and insert a No. 58 or No. 11 filter.

3. While looking through the focusing eyepiece, position the microscope slide so that the structures you would like to photograph are included in the correct reticle frame. Turn the eyepiece to include all the structures of interest and to obtain the best composition.

4. Turn the camera back so that it is at the same angle as the reticle frame.

5. Turn the collar of the focusing eyepiece until the cross hairs on the reticle are in sharp focus. Then, while looking through the focusing eyepiece, bring the object into sharp focus with the fine adjustment knob. If eyeglasses are routinely worn, they should be left on for the entire focusing procedure. At this point, depending on the design of your photomicrographic assembly, you may have to divert the light from the focusing eyepiece to the camera.

6. Set the exposure meter for the film's ISO rating (50) and, if required, for the film format (35mm) and ocular magnification. Record the meter reading and all other information about the photomicrograph on Data Sheet 21.1.

7. Make an exposure series, bracketing the indicated exposure time. Each successive film frame should have twice the exposure of the preceding one, with the middle frame of the series having the indicated exposure time. So, for

[1]Polaroid is a trademark of Polaroid Corporation, Cambridge, MA.
[2]PolaPan is a trademark of Polaroid Corporation, Cambridge, MA.

example, if the exposure meter indicates 1/15 sec, a series of three frames could be used with exposure times of 1/30, 1/15, and 1/8 sec. Be sure to advance the film after each exposure and to record the exposure time for each frame on Data Sheet 21.1.

8. Photograph the stage micrometer scale so that you will be able to determine the exact magnification of the print.

PROCESSING TECHNICAL PAN FILM

Photographic chemicals should be handled with care; they are toxic if swallowed or inhaled, and repeated contact may cause skin irritation and allergic skin reaction. Always wipe up spills immediately, as neatness and cleanliness are essential to quality photomicrographs. The developer should be at 20° C and the other solutions and rinses at 18.5°–21° C. The diluted developer is discarded after use, but the stop bath and fixer can be reused.

1. Rewind the exposed film and remove the magazine from the camera.
2. In *total darkness*, and handling the film only by the edges and leader, load the exposed film on the developing reel. Place the loaded reel in the developing tank and then screw on the cover. The closed developing tank has openings in the cover for adding and pouring out liquids, but is lightproof, so that the processing can be done with the room lights on.
3. Make a 1:20 dilution of the *HC-110* developer stock solution and, using a water bath, adjust the temperature of the diluted stock solution to 20° C. As 300 ml will usually cover one developing reel, mix 15 ml of the HC-110 stock solution with 285 ml of water in a 600-ml beaker.
4. Add the diluted HC-110 stock solution to the developing tank and immediately rotate the reel back and forth several times with the agitator rod to dislodge air bubbles. Develop for 8 min, agitating the solution by rotating the reel for 5 sec every 30 sec, first clockwise and then counterclockwise. After 8 min, discard the developer by inverting the closed developing tank over the sink drain.
5. Immediately add about 300 ml of stop bath to the developing tank. Agitate for 15–30 sec and then return the stop bath to its original container.
6. Add about 300 ml of fixer to the developing tank. Fix for 2–4 min with frequent agitation and then return the fixer to its original container.
7. Open the developing tank and wash the film under running tap water for 5–15 min. The stream of water should run into the center of the reel within the tank. Wash time is reduced to only 1 min (and water conserved) if the film is first soaked in a solution of Kodak *Hypo Clearing Agent*.
8. Pour off the water and add about 300 ml of a solution of Kodak *Photo Flo* to the developing tank. A 30-sec soak in this wetting solution will prevent the formation of water spots on the film as it dries.
9. Remove the reel from the tank, separate the two parts of the reel, attach film clips to the ends of the film, and hang the film in a dust-free area to dry. When completely dry, cut the film into lengths of five or six frames and store in glassine envelopes.

PRINTING

Photographic chemicals should be handled with care; they are toxic if swallowed or inhaled, and repeated contact may cause skin irritation and allergic skin reaction. Always wipe up spills immediately, as neatness and cleanliness are essential to quality photomicrographs.

The developer should be at 20° C and the other solutions and rinses at 18.5°–24° C.

The diluted developer, stop bath, and fixer can be used for multiple prints during the same printing session. Always make certain that before the room light or enlarger lamp is turned on, unused print paper is in its original package or in a lightproof cabinet.

1. Turn on the safelight. The print paper can be handled with an OC safelight filter in a lamp with a 15-watt bulb kept at least 1.2 m (4 feet) from the paper. The print paper should not be exposed to the safelight (at 1.2 m) for more than 3 min.

2. With the room lights on, add the diluted developer (1 vol *Dektol* + 2 vol water), stop bath, and fixer to three trays.

3. Select the best negative in the exposure series (see p. 279).

4. Lightly brush any dust off the negative, using a camel's hair brush. Then place the negative, emulsion side down, in the negative carrier of the enlarger.

5. Turn off the room lights and place a piece of scrap print paper in the enlarger easel, emulsion side up.

6. Turn on the enlarger lamp and open the lens aperture all the way (lowest *f*-stop).

7. Using the height control knob, adjust the enlarger height above the easel until you obtain the approximate degree of enlargement desired. You will have to adjust the focus each time the enlarger height is changed. This is done by changing the length of the bellows. Compose the print by moving the entire easel and its borders. When you are satisfied with the composition, make the final adjustment in focus. Use an enlarger focusing scope to see when the grains of the negative's emulsion are in sharp focus on the scrap paper. When the image is in focus, set the lens aperture to *f*/8 and turn off the enlarger lamp.

8. Insert a **PC 2** filter in the filter drawer of the enlarger.

9. Place a fresh piece of print paper (cut to size) in the easel and place a projection print scale on top of the paper.

10. Turn on the enlarger lamp and expose the print paper for 1 min.

11. Develop the exposed print paper by immersing it, emulsion side down, into the developer and then immediately turning it over. Develop for 1 min with continuous agitation. The print should be handled with tongs, *not* fingers.

12. Drain the print by touching it to the upper edge of the tray and then transfer it to the stop bath. Agitate the print in the stop bath for 5 sec.

13. Drain and transfer the print to the fixer. Fix for 2 min with frequent agitation. After fixing, you can turn on the room lights.

14. Wash the print in running tap water for 4 min.

15. Examine the print and select the sector with the best exposure. The number printed in this sector is the approximate number of seconds (at *f*/8) required to obtain a print with that exposure. For the final print, you will want to use an exposure time between 10 and 15 sec. The lens aperture may have to be opened wider or stopped down, and the exposure time changed accordingly.

16. Make the final print, as earlier, but with the exposure time (and appropriate *f*-stop) that you have determined to be best. It is a common practice to make several prints, using a slightly different exposure time for each. An efficient way to make two or three print exposures is to make *all* the exposures first and then develop the several prints at the same time. When making each exposure, be sure that the other pieces of print paper are in a lightproof cabinet. When developing the several pieces of exposed print paper, continually interleaf them in the developing tray so that air bubbles will not be trapped next to the emulsion. To determine the exact magnification of the photomicrograph, print the negative of the stage micrometer scale. If you want to make prints with higher or lower contrast, replace the **PC 2** filter with a **PC 3** or **PC 1** filter, respectively, and make a new test exposure using the projection print scale. Be sure that the enlarger lamp and safelight are turned off when you have finished.

17. After the prints have been washed, wipe the surface with a squeegee and allow to air-dry at room temperature. The dry prints should be trimmed with a paper cutter and mounted on lightweight cardboard with *mounting tissue,* which requires a photographic tacking iron to apply.

PHOTOGRAPHING CELL PREPARATIONS WITH POLAROID TYPE 667 FILM

The Polaroid developing process uses a caustic jelly, which is spread between the print surface and the black backing. If any should accidentally get on your skin, immediately wipe it off and wash the area with plenty of water. Keep the jelly away from eyes, mouth, and clothing.

1. Load the film pack in the Polaroid camera back, following the instructions that accompany the film. Pull out the black paper shield.
2. Place the microscope slide on the stage and adjust the microscope for Köhler illumination (see Project 1, p. 7). After you have located the field you want to photograph, adjust the condenser iris for optimum contrast and insert a No. 58 or No. 11 filter in the light path.
3. While looking through the focusing eyepiece, position the microscope slide so that the structures you would like to photograph are included in the correct reticle frame. Turn the eyepiece to include all the structures of interest and to obtain the best composition.
4. Turn the camera back so that it is at the same angle as the reticle frame.
5. Turn the collar of the focusing eyepiece until the cross hairs on the reticle are in sharp focus. Then, while looking through the focusing eyepiece, bring the object into sharp focus with the fine adjustment knob. If eyeglasses are routinely worn, they should be left on for the entire focusing procedure. At this point, depending on the design of your photomicrographic assembly, you may have to divert the light from the focusing eyepiece to the camera.
6. Set the exposure meter for the film's ISO rating (3000) and, if required, for the film format (8.3 × 10.8 cm) and ocular magnification. Record the meter reading and all other information about the photomicrograph on Data Sheet 21.1.
7. Make an exposure series. If there is a dark slide close to the film plane of the camera, a total of five exposures can be made on the same print. First, pull out the dark slide as far as it will go. Expose the film for one-fourth of the indicated exposure time and then push in the dark slide about one-fifth of the way. Again, expose the film for one-fourth of the indicated exposure time and push in the dark slide another one-fifth of its travel. Repeat this procedure with one-half of the indicated exposure time, with the indicated exposure time, and with twice the indicated exposure time. Had the meter reading been 1/15 sec, the one print would now have five sections with exposure times of 1/60, 1/30, 1/15, ~1/8, and ~1/4 sec. The total exposure time in any section is the sum of all the exposure times received prior to covering that section with the dark slide. Make a note of the exposure times.
8. While holding the camera back securely, pull the white tab out of the camera, after which a yellow tab will appear. Immediately (again, holding the camera securely), pull the yellow tab *straight* out of the camera, using a steady, uninterrupted pull at moderate speed. After it is all the way out, note the time, as development has begun. After the recommended development time, which is 45 sec at 21° C (70° F) and 30 sec at 24° C (75° F) or above, quickly separate the print from the black backing, starting at a corner of the print at the end nearer the yellow tab. Fold the black backing, moist side in, and discard it.

9. Examine the print and select the section with the best exposure (the longer the exposure, the lighter the print). Then expose the next sheet of film, using the best exposure time. Record the exposure time on Data Sheet 21.1.

PHOTOGRAPHING CELL PREPARATIONS WITH POLAPAN FILM

1. Load the film in the 35mm camera back and advance the film three frames. After the camera is loaded and mounted, pull out the dark slide.
2. Place the microscope slide on the stage and adjust the microscope for Köhler illumination (see Project 1, p. 7). After you have located the field you want to photograph, adjust the condenser iris for optimum contrast and insert a No. 58 or No. 11 filter.
3. While looking through the focusing eyepiece, position the microscope slide so that the structures you would like to photograph are included in the correct reticle frame. Turn the eyepiece to include all the structures of interest and to obtain the best composition.
4. Turn the camera back so that it is at the same angle as the reticle frame.
5. Turn the collar of the focusing eyepiece until the cross hairs on the reticle are in sharp focus. Then, while looking through the focusing eyepiece, bring the object into sharp focus with the fine adjustment knob. If eyeglasses are routinely worn, they should be left on for the entire focusing procedure. At this point, depending on the design of your photomicrographic assembly, you may have to divert the light from the focusing eyepiece to the camera.
6. Set the exposure meter for the film's ISO rating (80) and, if required, for the film format (35mm) and ocular magnification. Record the meter reading and all other information about the photomicrograph on Data Sheet 21.1.
7. Make an exposure series, bracketing the indicated exposure time. Each successive film frame should have twice the exposure of the preceding one, with the middle frame of the series having the indicated exposure time. So, for example, if the exposure meter indicates 1/15 sec, a series of three frames could be used with exposure times of 1/30, 1/15, and 1/8 sec. Be sure to advance the film after each exposure and to record the exposure time for each frame on Data Sheet 21.1.

PROCESSING POLAPAN FILM

The processing pack contains a caustic jelly. If any should accidentally leak out, keep it away from eyes, mouth, and clothing. If any gets on your skin, immediately wipe it off and wash the area with plenty of water. Immediately after development, the processing pack should be discarded in the original box.

1. Rewind the exposed film and remove the cartridge from the camera.
2. Load the film and processing pack into the AutoProcessor, following the instructions that accompany the unit.
3. Close the cover of the AutoProcessor, making certain that it latches.
4. Grasp the lever at the side of the processor and flip it down at a steady, medium rate of speed.
5. Wait 5 sec and then turn the crank *fast*, at about *two* revolutions per second, in a *clockwise* direction. (Turning it counterclockwise will damage the unit.) As soon as the gear clicking is no longer heard, immediately stop turning the crank.
6. Wait 60 sec. This is the correct processing time for this film at 21° C (70° F). At lower temperatures, the processing time must be increased (see film instructions).

7. Grasp the lever and flip it up at a steady, medium rate of speed.

8. Turn the crank *as fast as possible*, in a clockwise direction (*not counterclockwise*). When you no longer hear the gear clicking, continue turning the crank for an additional three or four revolutions. The fast rewind minimizes residue on the film.

9. Open the processor cover. Remove and discard the processing pack and then remove the film cartridge. The film is dry and can immediately be cut and mounted using the Polaroid *35mm Slide Mounter* and Polaroid *35mm Slide Mounts*. The mounted transparencies are then ready for projection.

REFERENCES

Birnbaum, H. C. 1986. *Black-and-White Darkroom Techniques*. Kodak Publication KW-15, Eastman Kodak Co., Rochester, NY.

Delly, J. G. 1988. *Photography Through the Microscope*, 9th ed. Kodak Publication No. P-2. Eastman Kodak Co., Rochester, NY.

Lefkowitz, L. 1985. *Polaroid 35mm Instant Slide System: A User's Manual*. Focal Press, Boston.

Loveland, R. P. 1970. *Photomicrography: A Complete Treatise*, Vols. 1 and 2. John Wiley, New York.

Data Sheet 21.1

Checklist of important variables in photomicrography.

Film, ISO _____

Frame	Slide no., stage coord.	Tissue, stain	Trans. setting	Objective, ocular	Filters	Meter reading	Exp. time	Other info.
1								
2								
3								
4								
5								
6								
7								
8								
9								
10								
11								
12								
13								
14								
15								
16								
17								
18								
19								
20								

INSTRUCTOR'S GUIDE

This section lists the equipment, supplies, and reagents required for each project. Quantities are calculated for a class of 20 students working individually or in teams, as noted.

Note to the preparator: Most of the required reagents are toxic to some degree. They should never be pipetted by mouth, and any spills on skin or clothing should be washed off immediately with water.

PROJECT **1**

Equipment

20 compound microscopes, each equipped with an ocular micrometer, green filter, and $10\times$, $40\times$, and $100\times$ objectives

Phase-contrast microscopes (optional)

Supplies

Micrometer slides

Microscope slides

Coverslips, 22 mm², No. 1½

10 single-edge razor blades

5 fresh sprigs of *Elodea*

1 onion

3 droppers

10 pairs pointed forceps

Toothpicks

Kimwipes

1 culture of *Euglena gracilis*. The culture should last for at least 1–2 weeks if kept in a well-illuminated area (but out of direct sunlight) with the temperature between 18°–22° C. For long-term maintenance of the culture, use split pea medium: boil 40 yellow split peas in 1 liter of distilled water for 10 min. Allow the medium to cool and then inoculate immediately.

20 prepared slides of *Zea* stem, cross section

20 prepared slides of common bacterial forms (cocci, bacilli, spirilla)

Lens paper, enough for the entire semester

Reagents

100 ml distilled water dispensed in five dropping bottles

100 ml Lugol's iodine solution dispensed in five dropping bottles: dissolve 6.0 g KI and 4.0 g I in 100 ml distilled water.

5 bottles of Protoslo (Carolina Biological Supply Co.)

Immersion oil, enough for the entire semester

PROJECT **2**

Each student should be provided with two unstained, fixed blood smears, prepared any time up to 6 hr before class. The blood should come from a healthy individual other than a student. Care should be taken when handling blood since there is always the slight possibility that the blood is contaminated with infectious agents. Accordingly, disposable gloves should be worn, spills wiped up, the area disinfected, and waste materials disposed of properly. It is advisable that the individual make the smears using his/her own blood. For making the fixed blood smears, follow this procedure:

1. Clean the slides thoroughly. New slides from the box should always be degreased by dipping in 95% ethanol and wiping with a lint-free tissue, like Kimwipes.
2. Use aseptic technique to obtain the blood sample. After washing the hands with soap and water, and drying them, clean the tip of the ring or middle finger with a sterile alcohol swab. When the fingertip has dried, lance it with a sterile lancet, squeeze the finger, and wipe away the first drop of blood. *The lancet should never be used more than once and is to be discarded immediately.*
3. Again, squeeze the finger and place a drop of blood near the frosted end of one slide. Hold a second slide with its edge at about a 30° angle on the first and bring it toward the drop; see (a) in the accompanying figure. After contact is made and the drop spreads along the edge of the slide (b), push the second slide to the other end in one smooth motion (c). In the preparation of multiple smears, the slide used to push should be the next one on which the smear is made. After the smears have been made, clean the lanced fingertip with alcohol.
4. Allow the smears to air dry.
5. Fix the air-dried smears in a 9:1 mixture of absolute ethanol:formalin for 10 min, in a Coplin jar.
6. Pour off and discard the fixative, using forceps, not fingers, to restrain the slides.
7. Rinse the slides in the Coplin jar under running tap water for 2 min. Then replace with distilled water and pour it off.
8. Remove the slides from the Coplin jar and allow them to air dry.

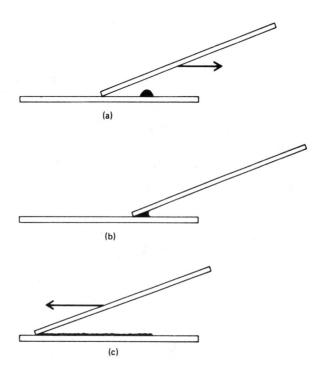

(a)

(b)

(c)

The steps (a), (b), and (c) for preparing a thin blood smear. See text for explanation.

As a check on whether the buffer alone removes PAS-positive material, one or several students can place the slide without diastase treatment in buffer at 24° C. As a check on whether preexisting aldehydes are being stained by Schiff's reagent, one or several students can be assigned an additional control slide that is not subjected to periodic acid treatment.

Equipment
20 compound microscopes

Supplies
Microscope slides with frosted end   Sterile alcohol swabs
Coverslips, 22 × 40 mm   Sterile blood lancets
  2 10-ml graduated cylinders   Parafilm
  2 250-ml graduated cylinders   Kimwipes
20 pairs forceps   Benchkote
Draining slide holders
20 Wright- or Giemsa-stained human blood smears. They can be purchased from Carolina Biological Supply Co. or Ward's Natural Science Establishment, Inc. Alternatively, they can be prepared simply and quickly using VWR Stat Stain (available from VWR Scientific).

Reagents
40 ml 95% ethanol in a Coplin jar

200 ml 9:1 ethanol:formalin fixative dispensed in five Coplin jars: mix 180 ml absolute ethanol and 20 ml formalin (37% w/w formaldehyde). Prepare shortly before use.

120 ml 1% diastase solution dispensed in three Coplin jars: dissolve 1.20 g diastase of malt (Fisher Scientific) in 120 ml M/15 phosphate buffer, pH 6.0. Buffer: dissolve 3.81 g KH_2PO_4, 0.76 g Na_2HPO_4 in distilled water to make 500 ml of solution. The diastase solution is good for 1 week if refrigerated.

300 ml 1% periodic acid dispensed in eight Coplin jars: dissolve 3.0 g H_5IO_6 in 300 ml distilled water. The solution is good for several weeks if stored in a dark bottle.

100 ml 1 N HCl: in a fume hood, dilute 8.1 ml concentrated HCl to 100 ml with distilled water.

220 ml Schiff's reagent (Sigma Chemical Co., cat. no. S5133) dispensed in seven Coplin jars

100 ml 10% metabisulfite solution: dissolve 10 g Na or K metabisulfite in distilled water to make 100 ml of solution.

75 ml Harris's hematoxylin (any commercial brand) dispensed in five Coplin jars. Shake well before dispensing.

175 ml xylene dispensed in five Coplin jars

50 ml Permount dispensed in five glass-rod dropping bottles

PROJECT **3**

Each student should be provided with two unstained, fixed blood smears, prepared any time up to 6 hr before class. The blood should come from a healthy individual other than a student. Care should be taken when handling blood since there is always the slight possibility that the blood is contaminated with infectious agents. Accordingly, disposable gloves should be worn, spills wiped up, the area disinfected, and waste materials disposed of properly. It is advisable that the individual make the smears using his/her own blood. For making the fixed blood smears, follow the procedure on page 292, except that here (1) use a larger drop of blood so that the smear will be thicker, (2) use Carnoy's fixative, a 6:3:1 mixture of absolute ethanol:chloroform:glacial acetic acid, and (3) dip the slides in 95% ethanol for several seconds before rinsing with tap water.

As a check on whether the distilled water alone removes RNA, one or several students can place the slide without RNase treatment in distilled water at 37° C.

Equipment

20 compound microscopes	1 37° C water bath

Supplies

Microscope slides with frosted end	Sterile alcohol swabs
Coverslips, 22 × 40 mm	Sterile blood lancets
20 pairs forceps	Kimwipes
Benchkote	Draining slide holders

Reagents

40 ml 95% ethanol in a Coplin jar

200 ml Carnoy's fixative dispensed in five Coplin jars: mix 120 ml absolute ethanol, 60 ml chloroform, and 20 ml glacial acetic acid. Prepare shortly before use.

120 ml 0.1% RNase solution dispensed in three Coplin jars: dissolve 120 mg RNase (Sigma Chemical Co., cat. no. R4875) in 120 ml distilled water. Adjust the pH to 6.5–7.0 with 0.1 N NaOH. Solution is good for 1 week if refrigerated.

175 ml buffered methyl green–pyronin staining solution (method of Perry and Reynolds) dispensed in five Coplin jars: dissolve 1.0 g of the *pure dye* methyl green (Harleco) in 200 ml hot 0.1 M acetate buffer, pH 4.4. For example, if a batch of methyl green has a dye content given as 89% on the label, add 1.12 g. When cool, transfer to a separatory funnel and extract repeatedly (in a fume hood) with 30-ml aliquots of chloroform until the chloroform is colorless or nearly so. The chloroform removes the contaminant, methyl violet. Add 0.2 g of the *pure dye* pyronin Y (Kodak Laboratory Chemicals) to 200 ml of the chloroform-extracted methyl green solution. For example, with a batch of pyronin Y that has a dye content of 50% and with 190 ml of chloroform-extracted methyl green solution, add 0.38 g. Shake well and dispense. The staining solution, which should be stored in a dark stoppered bottle, can be reused and is good for one week.

200 ml acetate buffer, pH 4.4. Make up 1 liter of 0.1 N acetic acid (5.7 ml glacial acetic acid diluted to 1 liter with distilled water) and 500 ml 0.1 M sodium acetate (4.10 g sodium acetate, anhydrous, dissolved in distilled water to make 500 ml of solution). Mix 124 ml of the 0.1 N acetic acid solution and 76 ml of the 0.1 M sodium acetate solution. Adjust pH to 4.4 with the acetic acid or acetate solution.

200 ml 3:1 *tert*-butanol:ethanol dispensed in five Coplin jars: mix 150 ml *tert*-butanol and 50 ml absolute ethanol.

350 ml xylene dispensed in 10 Coplin jars

50 ml Permount dispensed in five glass-rod dropping bottles

PROJECT **4**

Students should be provided with fixed root tips that are ready for staining. *Vicia faba* seeds need to be germinated about 2 weeks before the project will be done. Directions for germination and fixation are given under Supplies. For processing the root tips for the Feulgen reaction, students should work in teams of four, but for making the squashes, students should work individually. Quantities are calculated for 20 students working in five teams. The project can be carried out without counterstaining and without making the preparations permanent. The coverslips of temporary squashes should be sealed with Vaseline. If the mitotic stages are to be photographed in a future lab session, permanent squash preparations must be made.

Equipment

20 compound microscopes	3 60°C water baths
1 90°C water bath	

Supplies

Microscope slides with frosted end	5 50-ml beakers
Coverslips, 22 mm^2	Kimwipes
2 250-ml graduated cylinders	Aluminum foil
20 single-edge razor blades	Parafilm
20 pairs forceps, stainless steel	Benchkote
20 glass rods with fire-polished, flat ends	

50 fixed root tips of *Vicia faba*. Soak the seeds in a 10% aqueous solution of Clorox (or comparable hypochlorite product) for 10 min, rinse well in running water, and then plant in sterile vermiculite. Water the vermiculite initially and every few days thereafter. When the primary root is 5–8 cm long (about a week after planting), cut off the root tip and replace the seedling in the vermiculite. In about a week there should be many lateral roots. The terminal 4–7 mm of the lateral roots should be fixed overnight in freshly prepared 1:3 acetic:ethanol (1 vol glacial acetic acid:3 vol absolute ethanol) and then transferred to 70% ethanol, in which it can be stored for several months at 4° C. Seeds of "bush fava beans" are available from W. Atlee Burpee Co., 300 Park Ave., Warminster, PA 18991.

Vaseline petroleum jelly and toothpicks. (These items are not required if permanent squash preparations will be made.)

Reagents

200 ml 1 N HCl dispensed in ten 50-ml beakers (for hydrolysis); 50 ml 1 N HCl (for bisulfite bleaching solution): in a fume hood, dilute 20.3 ml concentrated HCl to 250 ml with distilled water.

100 ml Schiff's reagent (Sigma Chemical Co., cat. no. S5133)

100 ml 45% (v/v) glacial acetic acid dispensed in five dropping bottles

100 ml 10% metabisulfite solution: dissolve 10 g Na or K metabisulfite in distilled water to make 100 ml of solution.

150 ml 5% (w/v) trichloroacetic acid dispensed in five beakers

For Permanent Preparations and Counterstaining

Dry ice

320 ml 95% ethanol dispensed in eight Coplin jars

320 ml absolute ethanol dispensed in eight Coplin jars

80 ml 0.5% (w/v) fast green FCF in 95% ethanol, dispensed in two Coplin jars

50 ml Euparal dispensed in five glass-rod dropping bottles

PROJECT **5**

Students should work in pairs; quantities are calculated for 20 students working in 10 teams. The homogenization should be done by one designated team. To confirm that the absorbance of DNA with the orcinol reaction is about 10% the absorbance of the same concentration of RNA, one or several teams can run the orcinol reaction with a DNA stock solution.

Equipment

UV spectrophotometer(s)

2 clinical centrifuges

1 Waring blender, jar thoroughly chilled

2 90° C water baths

10 hot plates

1 balance

10 spectrophotometers. *Note:* For measurements above 600 nm, the Milton Roy Spectronic 20 should be equipped with either the wide-range (400–700 nm) or red-sensitive (600–950 nm) phototube and filter.

Supplies

120 glass cuvettes

30 quartz cuvettes

30 g calf liver, frozen

10 15-ml centrifuge tubes

10 10-ml graduated cylinders

50 1-ml pipets

60 2-ml pipets

20 5-ml pipets

10 pipet fillers, rubber bulb type

130 test tubes, 16 × 125 mm, with screw caps

Pasteur pipets, 146 mm, with dropper bulbs

20 400-ml beakers for boiling-water and ice-water baths

10 test tube racks

10 permanent marking pens

Parafilm

Kimwipes

20 pairs protective goggles

Boiling chips

10 French curve templates and rulers

10 10-ml pipets

Reagents

100 ml 0.15 M NaCl–0.015 M Na_3 citrate, also referred to as standard saline citrate or SSC: dissolve 4.38 g NaCl, 2.20 g Na_3 citrate · $2H_2O$ in distilled water to make 500 ml of solution. Adjust the pH to 7.0 with 1 N HCl.

100 ml DNA stock solution, 100 μg/ml in SSC: dissolve 10 mg DNA (e.g., calf thymus DNA, Sigma Chemical Co., cat. no. D1501) in 100 ml of the saline–sodium citrate solution at room temperature. Check the absorbance at 260 nm and, if necessary, adjust the DNA concentration so that $A \cong 1$.

100 ml RNA stock solution, 50 μg/ml in SSC: dissolve 5 mg RNA (e.g., calf liver RNA, Sigma Chemical Co., cat. no. R7250) in 100 ml of the saline–sodium citrate solution at room temperature. Check the absorbance at 260 nm and, if necessary, adjust the RNA concentration so that $A \cong 1$.

120 ml ice-cold distilled water

150 ml ice-cold 10% (w/v) trichloroacetic acid (TCA) dispensed in 10 capped Erlenmyer flasks

350 ml 5% (w/v) TCA dispensed in 10 capped Erlenmyer flasks

250 ml 95% ethanol dispensed in 10 capped Erlenmyer flasks

100 ml each of four DNA stock solutions: 400 μg/ml, 300 μg/ml, 200 μg/ml, 100 μg/ml, dispensed in capped test tubes. To make the 400 μg/ml stock solution, add 100 mg DNA (e.g., calf thymus DNA, Sigma Chemical Co., cat. no. D1501) to 250 ml of 5% TCA. To dissolve, place the mixture in a 90° C water bath for several minutes, swirling occasionally. To make the 300 μg/ml stock solution, dilute 75 ml of the 400 μg/ml stock solution to 100 ml with 5% TCA. To make the 200 μg/ml stock solution, dilute 50 ml of the 400 μg/ml stock solution to 100 ml with 5% TCA. To make the 100 μg/ml stock solution, dilute 25 ml of the 400 μg/ml stock solution to 100 ml with 5% TCA.

100 ml each of four RNA stock solutions: 400 μg/ml, 300 μg/ml, 200 μg/ml, 100 μg/ml, dispensed in capped test tubes. Follow the same procedure used for making the four DNA stock solutions but, instead of DNA, use calf liver RNA (Sigma Chemical Co., cat. no. R7250).

300 ml diphenylamine reagent dispensed in 50-ml burets: in a fume hood, dissolve 3.0 g fresh or recrystallized (in petroleum ether) diphenylamine in 300 ml glacial acetic acid and 8.25 ml concentrated sulfuric acid. Prepare within 30 min before use.

250 ml orcinol reagent dispensed in 50-ml burets: in a fume hood, dissolve 2.5 g orcinol in 250 ml concentrated HCl containing 1.25 g $FeCl_3 \cdot 6H_2O$. Prepare within 30 min before use.

PROJECT 6

Students should work in teams of four; quantities are calculated for 20 students working in five teams. Each electrophoresis chamber accommodates up to three cellulose acetate plates, so that two electrophoresis chambers should be ample for an entire class.

The hemoglobin samples should be handled with care since there is always the slight possibility that the blood is contaminated with infectious agents. Accordingly, disposable gloves should be worn by students and others handling the hemoglobin samples, spills wiped up, the area disinfected, and waste materials disposed of properly. *The electrophoresis equipment should always be supervised because of the potential shock hazard associated with the power supply.*

The procedures assume the use of Titan III-H Cellulose Acetate Plates, the Super Z Applicator Kit, and Titan Gel Chamber, available from Helena Laboratories, 1530 Lindbergh Drive, P.O. Box 752, Beaumont, Texas 77704-0752. The trade names of all Helena electrophoresis items (including the registered trademarks *Titan*, *Supre Heme*, and *Zip Zone*) and catalog numbers are also listed.

Equipment

2 Titan Gel Chambers (4063) 1 56° C oven
2 power supplies
5 adjustable micropipetters, 2–10 μl (e.g., Eppendorf Digital Pipette)

Supplies

1 box 25 Titan III-H Cellulose Acetate Plates, 60 × 76 mm (3022)
5 Super Z Applicator Kits, 8 sample (4088). Each kit contains 1 eight-sample applicator, 2 eight-sample well plates, 1 aligning base, 1 bottle Zip Zone Prep detergent.
1 Bufferizer (containers for soaking plates) (5093)
1 700 Staining Set (5114). Contains 1 carrying rack for plates and 7 700-ml plastic containers with lids.
1 box 100 Titan Blotter Pads, 89 × 108 mm (5037)
1 box 500 Disposable Wicks (5081)
1 pkg 200 Titan Plastic Envelopes, 121 × 64 mm (5052)
Pipet tips
Disposable gloves
1 100-ml graduated cylinder
5 permanent marking pens
Benchkote

Reagents

1 box of 10 pkg Supre Heme Buffer (5802). Dissolve the contents of one package in 980 ml distilled water. The buffer in the electrophoresis chamber should be discarded after use. Buffer used to soak the cellulose acetate plates can be reused for soaking up to 12 plates.

AA_2 Hemo Control (5328)

ASA_2 Hemo Control (5329)

AFSC Hemo Control (5331)

5 250-ml bottles ponceau S stain (5525)

2 liters 5% acetic acid: bring 100 ml glacial acetic acid to 2 liters with distilled water.

PROJECT **7**

Students should work in teams of four; quantities are calculated for 20 students working in five teams. Each electrophoresis chamber accommodates up to three cellulose acetate plates, so that a total of two electrophoresis chambers should be ample for an entire class.

The serum samples should be handled with care since there is always the slight possibility that they are contaminated with infectious agents. Accordingly, disposable gloves should be worn by students and others handling the serum samples, spills wiped up, the area disinfected, and waste materials disposed of properly. *The electrophoresis equipment should always be supervised because of the potential shock hazard associated with the power supply.*

The procedures assume the use of Titan III Cellulose Acetate Plates, the Super Z Applicator Kit, and Titan Gel Chamber, available from Helena Laboratories, 1530 Lindbergh Drive, P.O. Box 752, Beaumont, Texas 77704-0752. The trade names of all Helena electrophoresis items (including the registered trademarks *Titan, Electra,* and *Zip Zone*) and catalog numbers are also listed.

Equipment

2 Titan Gel Chambers (4063)   1 56°C oven

2 power supplies

5 adjustable micropipetters, 2–10 µl (e.g., Eppendorf Digital Pipette)

1 densitometer (e.g., Quick Scan Jr. TLC, Helena cat. no. 1038)

Supplies

1 box 25 Titan III Cellulose Acetate Plates, 60 × 76 mm (3023)

5 Super Z Applicator Kits, 8 sample (4088). Each kit contains 1 eight-sample applicator, 2 eight-sample well plates, 1 aligning base, 1 bottle Zip Zone Prep detergent.

1 Bufferizer (containers for soaking plates) (5093)

1 700 Staining Set (5114). Contains 1 carrying rack for plates and 7 700-ml plastic containers with lids.

1 box 100 Titan Blotter Pads, 89 × 108 mm (5037)

1 box 500 Disposable Wicks (5081)

1 pkg 200 Titan Plastic Envelopes, 121 × 64 mm (5052)

Pipet tips

Disposable gloves

1 100-ml graduated cylinder

5 permanent marking pens

Benchkote

Reagents

1 box 10 pkg Electra HR Buffer (5805). Dissolve the contents of one package in 750 ml distilled water. The buffer in the electrophoresis chamber should be discarded after use. Buffer used to soak cellulose acetate plates can be reused for soaking up to 12 plates.

Sera: calf, 1 ml (Sigma Chemical Co., cat. no. S1632); goat, 1 ml (S2007); guinea pig, 1 ml (S3634); horse, 2 ml (S6380). Dissolve the lyophilized serum in the indicated volume of distilled water.

5 250-ml bottles of ponceau S stain (5525)

2 liters 5% acetic acid: bring 100 ml glacial acetic acid to 2 liters with distilled water.

1 liter methanol

500 ml clearing solution: 150 ml glacial acetic acid, 350 ml methanol, 20 ml polyethylene glycol (Clear Aid, 5005)

PROJECT **8**

Students should work in teams of four; quantities are calculated for 20 students working in five teams. Each team is responsible for obtaining one of the tissue extracts and then dispensing the diluted extract to other members of the class. Each team can then electrophorese extracts from four different tissues. As an alternative to using extracts prepared in class, commercially prepared tissue extracts from rat heart and liver can be used (LD Isozyme Control, cat. no. 5919, Helena Laboratories). Each vial of lyophilized extract is reconstituted with 1.0 ml distilled water. The reconstituted extract is applied (3.0 µl) to the agarose gel plate *without dilution*. If this option is selected, be sure to follow all safety precautions and recommendations for storage given on the package insert.

Each electrophoresis chamber accommodates two agarose gel plates, so that three electrophoresis chambers should be ample for an entire class. *The electrophoresis equipment should always be supervised because of the potential shock hazard associated with the power supply.*

The procedures assume the use of the Titan Gel Isoenzyme Kit and Titan Gel Chamber, available from Helena Laboratories, 1530 Lindbergh Drive, P.O. Box 752, Beaumont, Texas 77704-0752. The trade names of all the Helena electrophoresis items (including the registered trademark *Titan*) and catalog numbers are also listed.

Electrophoresis Equipment

3 Titan Gel Chambers (4063)

3 power supplies

5 adjustable micropipetters, 2–10 µl (e.g., Eppendorf Digital Pipette)

1 45° C incubator

1 60° C oven

1 densitometer (optional)

Electrophoresis Supplies

1 Titan Gel Isoenzyme Kit (3043). Contains 10 agarose gel plates, 10 vials LDH isoenzyme reagent, diluent, buffer, templates, and blotters A, B, and D. The items supplied with the kit may change occasionally. Minor changes in Helena Laboratories' recommended procedures may occur; instructions are supplied with each kit.

5 Titan Gel Isoenzyme Incubation Chambers (4062)

1 box development slides (76 × 102 mm glass plates) (5008)

5 development weights (5014)

1 pkg 100 Titan Plastic Envelopes, 120 × 102 mm (5053)

Pipet tips

5 5-ml pipets

5 large finger bowls, 15–20 cm in diameter

1 50-ml graduated cylinder

5 permanent marking pens

Electrophoresis Reagents

LDH buffer, pH 8.1-8.3 (supplied with Titan Gel Isoenzyme Kit). Dissolve the contents of the package in 1500 ml distilled water. The buffer in the electrophoresis chamber should be discarded after use.

LDH isoenzyme reagent (NAD, lithium lactate, nitro blue tetrazolium, phenazine methosulfate; supplied with Titan Gel Isoenzyme Kit). Reconstitute the contents of each vial with 1.0 ml diluent (supplied) about 5 min before electrophoresis is completed. Invert each vial repeatedly until all material is dissolved. All the reconstituted reagent in each vial is used for one gel plate.

2 liters 10% acetic acid: bring 200 ml glacial acetic acid to 2 liters with distilled water.

Homogenization Equipment

5 10-ml Potter–Elvehjem tissue grinders

2–4 stirrer motors

1 balance

1 refrigerated centrifuge or a clinical centrifuge, prechilled and run in a cold room or refrigerator

Homogenization Supplies

1 rat

1 dissecting tray

Disposable gloves

5 wooden chopping blocks

5 single-edge razor blades

5 1-ml pipets

5 2-ml pipets

5 pipet fillers, rubber bulb type

10 centrifuge tubes

20 5-ml vials

2 50-ml graduated cylinders

5 50-ml Erlenmyer flasks

5 400-ml beakers for ice-water baths

Parafilm

Dissecting instruments: scalpel, scissors, forceps, dissecting pins

Homogenization Reagents

2 liters ice-cold distilled water

PROJECT

Equipment

20 compound microscopes equipped with ocular micrometers

Supplies

Microscope slides with frosted end

Coverslips, 22 mm^2, No. 1 ½

10 fresh sprigs of *Elodea*

20 pairs forceps

Kimwipes

Reagents

100 ml of each of the following sucrose solutions dispensed in five dropping bottles. Add the indicated amount of sucrose to a volumetric flask and bring the volume to 100 ml with distilled water: 0.2 M sucrose (6.85 g), 0.3 M sucrose (10.27 g), 0.4 M sucrose (13.69 g), 0.5 M sucrose (17.12 g), 0.6 M sucrose (20.54

g). Store at 4° C. Just before use at room temperature, swirl each dropping bottle.

100 ml of each of the following alcohol–sucrose solutions dispensed in five dropping bottles. Add the indicated volume of the alcohol to a volumetric flask and bring the volume to 100 ml with 0.3 M sucrose: 0.4 M methanol–0.3 M sucrose (1.6 ml), 0.4 M ethanol–0.3 M sucrose (2.3 ml), 0.4 M ethylene glycol–0.3 M sucrose (2.2 ml), 0.4 M 1-propanol–0.3 M sucrose (3.0 ml), 0.4 M propylene glycol–0.3 M sucrose (2.9 ml), 0.4 M glycerol–0.3 M sucrose (2.9 ml). Store at 4° C. Just before use at room temperature, swirl each dropping bottle.

PROJECT **10**

For the cell fractionation procedures, students should work in teams of four; quantities are calculated for 20 students working in five teams. For the microscopic examination of the fractions, students should work individually. The bulk of the project can be completed even if a refrigerated, high-speed centrifuge is not available since both the chloroplast fraction of spinach and the nuclear fraction of cauliflower can be isolated with a clinical centrifuge. For the nuclear fraction, the centrifuge should be prechilled and run in a cold room or refrigerator.

Equipment
1 clinical centrifuge with a swinging-bucket rotor
1 centrifuge tachometer
1 refrigerated, high-speed centrifuge
20 compound microscopes equipped with ocular micrometers

Supplies

10 mortars and pestles, chilled	Microscope slides
10 funnels	Coverslips, 22 mm^2
5 5-ml pipets	5 spatulas
5 25-ml graduated cylinders	5 pairs scissors
5 50-ml graduated cylinders	Parafilm
5 15-ml test tubes	Kimwipes
5 small test tubes, 10 × 75 mm	5 rulers
5 single-edge razor blades, wiped with alcohol to remove grease	
5 400-ml beakers for ice-water baths	
12 15-ml round bottom centrifuge tubes	
5 pipet fillers, rubber bulb type	

Pasteur pipets, 146 mm, and dropper bulbs

200 g fresh spinach. Just before use, rinse with cold tap water and blot dry.

2 fresh, firm heads of cauliflower, refrigerated until used. Just before use, rinse with cold tap water and blot dry.

12 50-ml centrifuge tubes for first centrifugation (for obtaining the nuclear fraction with a refrigerated, high-speed centrifuge). With a clinical centrifuge, 15-ml conical centrifuge tubes can be used.

6 50-ml or 10 25-ml centrifuge tubes for second centrifugation (for obtaining the mitochondrial fraction)

Cotton cheesecloth, grade 10 (Curity). Available from Thomas Scientific.

Reagents
50 g purified sea sand

250 ml 0.35 M NaCl–0.02 M Tris buffer, pH 7.5: dissolve 5.11 g NaCl, 0.61 g tris(hydroxymethyl) aminomethane in distilled water to make 250 ml of solution. Adjust pH to 7.5 with 6 N HCl. Store at 4° C.

100 ml of a saturated solution of urea in distilled water, dispensed in five dropping bottles

100 ml 2% (v/v) Triton X-100 in distilled water, dispensed in five dropping bottles

500 ml mannitol grinding medium, 0.3 M D-mannitol-0.02 M phosphate buffer, pH 7.2: dissolve 27.33 g D-mannitol, 0.41 g KH_2PO_4, 1.21 g K_2HPO_4 in distilled water to make 500 ml of solution. Adjust pH to 7.2 with 1 N KOH. Store at 4° C.

90 ml 1% lacto-aceto-orcein dispensed in five dropping bottles. Add 0.90 g natural orcein (Sigma Chemical Co.) to a mixture of 30 ml lactic acid and 30 ml glacial acetic acid in a 125-ml Erlenmyer flask. Stir for 20 min with a magnetic stirrer. Add 30 ml distilled water, mix, and dispense. If there are undissolved particles, filter through Whatman No. 1 filter paper with suction.

40 ml 0.025% (w/v) Janus green B dispensed in five dropping bottles: dilute 1.0 ml of a 1% (w/v) aqueous stock solution of the stain with 39 ml of the mannitol grinding medium. The stock solution is stable if stored in a dark, stoppered bottle. The working solution should be prepared shortly before use.

PROJECT **11**

Students should work in teams of four; quantities are calculated for 20 students working in five teams.

Equipment

5 spectrophotometers	2 balances
1 refrigerated, high-speed centrifuge	2 hot plates

Supplies

5 mortars and pestles, chilled	5 15-ml test tubes
5 funnels	50 cuvettes
5 50-ml graduated cylinders	5 test tube racks
25 1-ml pipets	5 spatulas
5 5-ml pipets	2 pairs scissors
5 10-ml pipets	5 permanent marking pens
5 pipet fillers, rubber bulb type	Parafilm
5 400-ml beakers for ice-water baths	Kimwipes
5 French curve templates and rulers	Pasteur pipets, 146 mm, and dropper
2 250-ml beakers for boiling-water baths	bulbs

5 single-edge razor blades, wiped with alcohol to remove grease

2 fresh, firm heads of cauliflower, refrigerated until used. Just before use, rinse with cold tap water and blot dry

6 50-ml centrifuge tubes for first centrifugation

6 50-ml or 10 25-ml centrifuge tubes for second centrifugation

Cotton cheesecloth, grade 10 (Curity). Available from Thomas Scientific.

Reagents

25 g purified sea sand

500 ml grinding medium, 0.3 M D-mannitol–0.02 M phosphate buffer, pH 7.2:

dissolve 27.33 g D-mannitol, 0.41 g KH_2PO_4, 1.21 g K_2HPO_4 in distilled water to make 500 ml of solution. Adjust pH to 7.2 with 1 N KOH. Store at 4° C.

500 ml assay medium, 0.3 M D-mannitol–0.02 M phosphate buffer–0.01 M KCl–0.005 M $MgCl_2$ dispensed in 10 Erlenmyer flasks: dissolve 27.33 g D-mannitol, 0.41 g KH_2PO_4, 1.21 g K_2HPO_4, 0.38 g KCl, 0.51 g $MgCl_2 \cdot 6H_2O$ in distilled water to make 500 ml of solution. Adjust pH to 7.2 with 1 N KOH. Store at 4° C.

50 ml 0.04 M sodium azide dispensed in five 15-ml test tubes: dissolve 0.26 g sodium azide* in distilled water to make 100 ml of solution.

50 ml 5×10^{-4} M DCIP dispensed in five 15-ml test tubes: dissolve 145 mg 2, 6-dichlorophenolindophenol sodium salt in distilled water to make 1000 ml of solution. Prepare shortly before use.

50 ml 0.2 M Na_2 malonate dispensed in five 15-ml test tubes: dissolve 3.32 g Na_2 malonate $\cdot H_2O$ in distilled water to make 100 ml of solution. Adjust pH to 7.0 with 1 N HCl.

50 ml 0.2 M Na_2 succinate dispensed in five 15-ml test tubes: dissolve 5.40 g Na_2 succinate $\cdot 6H_2O$ in distilled water to make 100 ml of solution. Adjust pH to 7.0 with 1 N HCl.

PROJECT **12**

Students should work in pairs; quantities are calculated for 20 students working in 10 teams. Since the petroleum ether–acetone solvent is extremely flammable, the chromatography should be carried out in a fume hood.

Equipment

10 spectrophotometers. See note on page 296.

10 chromatography jars. Lids should have an affixed hook for attaching the chromatography paper. Small chromatography jars give better results than large jars.

4 balances

Supplies

Whatman 3MM chromatography paper

10 mortars and pestles, chilled

10 25-ml graduated cylinders

5 250-ml graduated cylinders

10 50-ml test tubes with corks

50 cuvettes with corks

10 French curve templates and rulers

10 small test tubes, 10 × 75 mm

Pasteur pipets, 146 mm, and dropper bulbs

10 capillary tubes, length 8–10 cm, i.d. 1 mm. Melting-point capillaries work well.

10 5-ml pipets

10 pipet fillers, rubber bulb type

100 g fresh spinach

10 test tube racks

10 pairs scissors

10 pairs forceps

10 permanent marking pens

Vaseline petroleum jelly

Kimwipes

Reagents

30 g purified sea sand

2 liters petroleum ether, 35°–60° C boiling range

1 liter acetone; 200 ml should be ice-cold.

*CAUTION: *Sodium azide is poisonous and in the dry state is explosive if subjected to shocks.*

Students should work in pairs; quantities are calculated for 20 students working in 10 teams.

Equipment

10 spectrophotometers	4 balances
2 clinical centrifuges	

Supplies

10 mortars and pestles, chilled	100 cuvettes
5 25-ml graduated cylinders	10 250-ml beakers
10 400-ml beakers for ice-water baths	10 thermometers
10 15-ml test tubes	10 funnels
50 1-ml pipets	10 test tube racks
20 5-ml pipets	10 pairs scissors
10 10-ml pipets	10 permanent marking pens
10 pipet fillers, rubber bulb type	Aluminum foil
10 French curve templates and rulers	Parafilm
20 15-ml conical centrifuge tubes	Kimwipes

200 g fresh spinach. Just before use, rinse with cold tap water and blot dry.

Pasteur pipets, 146 mm, and dropper bulbs

10 gooseneck or table lamps with 100-watt frosted incandescent bulbs

Cotton cheesecloth, grade 10 (Curity). Available from Thomas Scientific.

Reagents

30 g purified sea sand

500 ml 0.35 M NaCl–0.02 M Tris buffer, pH7.5, dispensed in 10 Erlenmyer flasks: dissolve 20.45 g NaCl, 2.42 g tris (hydroxymethyl) aminomethane in distilled water to make 1000 ml of solution. Adjust pH to 7.5 with 6 N HCl. Store at 4° C.

100 ml 4×10^{-4} M DCIP dispensed in 10 15-ml test tubes: dissolve 116 mg 2,6-dichlorophenolindophenol sodium salt in distilled water to make 1000 ml of solution. Prepare shortly before use.

100 ml 0.01 N ammonia dispensed in 10 15-ml test tubes: in a fume hood, prepare a 1 N solution by bringing 6.77 ml concentrated (28%) NH_4OH to 100 ml with distilled water. Then bring 5.0 ml of the 1 N solution to 500 ml with distilled water.

100 ml 10^{-4} M DCMU dispensed in 10 15-ml test tubes: dissolve 24 mg 3-(3,4-dichlorophenyl)-1,1-dimethylurea (Sigma Chemical Co.) in distilled water to make 1000 ml of solution.

100 ml distilled water dispensed in 10 15-ml test tubes

In the experiment on flagellar function, the deflagellation, centrifugation, and dispensing of samples should be done by one designated person.

Equipment

4 dissecting microscopes	1 clinical centrifuge
1 pH meter	1 reciprocating shaker

1 magnetic stirrer and stirring bar

20 compound microscopes equipped with ocular micrometers

Supplies

Microscope slides

Coverslips, 22 mm², No. 1½

80 small test tubes, 10 × 75 mm

1 100-ml beaker

2 10-ml graduated cylinders

2 15-ml graduated, conical centrifuge tubes

2 50-ml Erlenmyer flasks with cotton plugs

Pasteur pipets, 146 mm, and dropper bulbs

1 fluorescent lamp with two cool-white, 15-watt bulbs

10 droppers

20 test tube racks

20 permanent marking pens

Vaseline petroleum jelly

Toothpicks

Kimwipes

Several cultures of *Pelomyxa carolinensis* and/or *Amoeba proteus*. *P. carolinensis* is preferred since they are larger. For short-term (7–10 days) maintenance of the cultures, add two previously boiled wheat grains to each culture as soon as they arrive from the biological supply company (assuming that no wheat or rice grains are already present in the culture). Long-term cultures can be maintained using the following medium: add three or four previously boiled wheat grains to 200 ml pasteurized (60° C for 15 min) spring water while still hot. Allow the medium to cool and then inoculate. For culturing *P. carolinensis*, also add *Paramecium caudatum*, on which the *P. carolinensis* feed.

Several cultures of *Paramecium caudatum*. For short-term (7–10 days) maintenance of the cultures, add two previously boiled wheat grains to each culture as soon as they arrive from the biological supply company (assuming that no wheat or rice grains are already present in the culture). Long-term cultures can be maintained using the following medium: add six to eight previously boiled wheat grains to 200 ml pasteurized (60° C for 15 min) spring water while still hot. Allow the medium to cool and then inoculate.

1 culture of *Chlamydomonas reinhardi*. For the experiment on flagellar function, the *Chlamydomonas* are cultured in Medium I of Sager and Granick (see Reagents), as follows. Using sterile technique, inoculate 75 ml of Medium I in a 250-ml Erlenmyer flask. The culture is grown at room temperature with constant aeration (cotton-filtered air) and constant illumination (fluorescent lamp). The culture is ready for use in the experiment when A_{675} is between 0.1 and 0.5, a culture density range that is reached 2–3 days after inoculation. Commercially available cultures of *Chlamydomonas* may be supplied growing on agar slants, on which the organisms lack flagella. Thus, even for the initial observations on *Chlamydomonas*, they must be grown in liquid Medium I for 2–3 days. Long-term cultures of *Chlamydomonas* are easily maintained on an agar slant of Medium I (10 ml Medium I + 0.15 g agar). The slant should be kept in a well-illuminated area at room temperature.

Reagents

10 ml suspension of Congo red–stained yeast: add 3 g compressed yeast and 30 mg Congo red to 10 ml distilled water; boil gently for 10 min.

100 ml Lugol's iodine solution dispensed in five dropping bottles: see page 291.

30 ml 0.5 N acetic acid dispensed in a dropping bottle: dilute 2.9 ml glacial acetic acid to 100 ml with distilled water.

30 ml 0.5 N KOH dispensed in a dropping bottle

150 ml Medium I of Sager and Granick: to a 1000-ml volumetric flask, add 0.5 g Na_3 citrate · $2H_2O$, 0.1 g K_2HPO_4, 0.1 g KH_2PO_4, 0.3 g NH_4NO_3, 0.3 g $MgSO_4$ · $7H_2O$, 0.04 g $CaCl_2$, 0.01 g $FeCl_3$ · $6H_2O$, 10 ml trace metal solution (see below),

distilled water to make 1000 ml of solution. Dissolve the sodium citrate first to avoid precipitation of any salts. The medium is autoclaved and then refrigerated until use. Medium I is also required for Project 15.

1000 ml trace metal solution: dissolve 100 mg H_3BO_3, 100 mg $ZnSO_4 \cdot 7H_2O$, 40 mg $MnSO_4 \cdot 4H_2O$, 20 mg $CoCl_2 \cdot 6H_2O$, 20 mg $Na_2MoO_4 \cdot 2H_2O$, 4 mg $CuSO_4$ in distilled water to make 1000 ml of solution.

5 bottles of Protoslo (Carolina Biological Supply Co.)

PROJECT **15**

Students should work in teams of three; quantities are calculated for 21 sutdents working in seven teams. The deflagellation procedure and dispensing of samples should be done by one designated team.

Equipment

1 pH meter
1 magnetic stirrer and stirring bar
21 compound microscopes equipped with ocular micrometers

1 clinical centrifuge
1 reciprocating shaker

Supplies

224 small test tubes, 10 × 75 mm
1 50-ml graduated cylinder
4 10-ml pipets
1 pipet filler, rubber bulb type
3 15-ml graduated conical centrifuge tubes
4 50-ml Erlenmyer flasks with cotton plugs
10 French curve templates and rulers

Microscope slides
Coverslips, 22 mm^2
1 100-ml beaker
21 test tube racks
21 permanent marking pens
Kimwipes

Pasteur pipets, 146 mm, and dropper bulbs

1 fluorescent lamp with two cool-white, 15-watt bulbs
1 culture of *Chlamydomonas reinhardi*. The culture procedure is the same as in the experiment on flagellar function, Project 14, page 305.

Reagents

150 ml Medium I; see page 305.

10 ml Medium I containing colchicine, 3 mg/ml: dissolve 30 mg colchicine in 10 ml Medium I.

10 ml Medium I containing cycloheximide, 10 μg/ml: dissolve 10 mg cycloheximide in 10 ml Medium I. Then dilute 0.5 ml of this solution to 50 ml with Medium I.

100 ml Lugol's iodine solution dispensed in five dropping bottles; see page 291.

30 ml 0.5 N acetic acid dispensed in a dropping bottle; see page 305.

30 ml 0.5 N KOH dispensed in a dropping bottle

Optional Exercises

1. To determine whether the effects of colchicine are reversible, the colchicine medium can be replaced with fresh Medium I (without colchicine) part way through the experiment. After the 20-min sampling, the colchicine culture is centrifuged for 5 min at 1300 g and the pellet resuspended in 10 ml fresh Medium I. The suspension is transferred to a clean 50-ml Erlenmyer flask, returned to the

shaker and sampled for the duration of the experiment. For this study, the initial concentration of colchicine should be 1.5 mg/ml.

2. To ascertain whether colchicine induces disassembly of flagellar microtubules, the colchicine can be added after some regeneration has occurred. Specifically, an aliquot of the deflagellated culture is allowed to undergo regeneration in Medium I for 30 min (just like the control culture). After the 30-min sampling, colchicine is added directly to the culture (final concentration = 3 mg/ml). Flagellar length is measured at the same sampling times used for the other cultures.

PROJECT **16**

Students should be provided with fixed, stained testes from last-instar or young, adult male grasshoppers. Students should work individually, though one testis can be shared by each pair of students. Directions for dissection, fixation, and staining are given under Supplies. A species commonly used for chromosome analysis in North America is *Melanoplus femur-rubrum*, although many other species are also suitable. Grasshoppers can be collected locally, with one collection trip providing enough material for several years' use. An excellent pictorial key is *How to Know the Grasshoppers, Cockroaches and Their Allies* by J. R. Helfer (Wm. C. Brown Co., Dubuque, IA, 1963).

For the study of spermiogenesis, ideally the preparations should be counterstained with fast green. This requires making the preparations permanent with the dry-ice method of Conger and Fairchild (Project 4, p. 33). The stages of meiosis can be studied in Vaseline-sealed squashes, though making them permanent allows photomicrographs to be taken in a future lab session. As an alternative to having students prepare the testis squashes, prepared slides can be supplied. Aceto-carmine squashes of grasshopper testis are available from Ward's Natural Science Establishment, Inc. (cat. no. 93W2249).

Equipment

20 dissecting microscopes	20 compound microscopes

Supplies

Microscope slides	20 Syracuse dishes
Coverslips, 22 mm^2	Kimwipes
20 pairs dissecting needles	

10 fixed, stained grasshopper testes. Dissection and fixation: in grasshoppers, the testes are fused into one large, yellow-orange body, which is exposed by tearing open the ventral abdomen with a pointed forceps. After removal, it is immediately fixed in freshly prepared 1:3 acetic:ethanol (1 vol glacial acetic acid:3 vol absolute ethanol). The dissection and fixation can be done in the field. Following overnight fixation, the testes are transferred to 70% ethanol, in which they can be stored almost indefinitely at 4° C. Staining: the entire testis is stained by placing it in alcoholic HCl–carmine (see Reagents) for 16 hr. After staining, the testis should be stored in 70% ethanol at 4° C.

Vaseline petroleum jelly and toothpicks. (These items are not required if permanent squash preparations will be made.)

Reagents

100 ml alcoholic HCl–carmine (method of R. Snow): dissolve 4.0 g carmine alum lake (Fisher Scientific) in 10 ml 1 N HCl + 20 ml distilled water in a 125-ml Erlenmyer flask. Plug the flask with cotton and boil gently for 10 min. When cool, add 70 ml absolute ethanol and then filter through coarse filter paper

with suction. The stain can be reused and will last almost indefinitely if stored in a dark, tightly stoppered bottle. It does not need to be filtered again.

200 ml 70% ethanol

200 ml 45% (v/v) glacial acetic acid dispensed in 10 dropping bottles

For Permanent Preparations and Counterstaining

Supply the materials listed on page 295, as well as 20 pairs forceps and 20 single-edge razor blades.

PROJECT **17**

This project requires two lab sessions to complete. During the first lab session, students prepare the digests, electrophorese the samples, and place the gel in the stain. The bands are visible in several hours, though overnight staining is recommended. Accordingly, the following morning, either a team member or the instructor/lab assistant should remove the gel from the stain and do the destaining. In the second lab session, students can examine the gel and collect/analyze their data. The gels will last months if stored in the refrigerator in a sealed plastic bag wrapped with aluminum foil. So that the enzymatic treatments and electrophoresis can be completed in a single lab session, the agarose gels must be cast beforehand. Instructions for casting the gel and setting up the electrophoresis chamber are given under Reagents.

Students should work in teams of four; quantities are calculated for 20 students working in five teams. The restriction enzymes and phage lambda DNA are available from Bethesda Research Laboratories (BRL), Gaithersburg, MD 20877. *The electrophoresis equipment should always be supervised because of the potential shock hazard associated with the power supply.*

Equipment

5 horizontal gel electrophoresis chambers with casting decks and well formers, available from E-C Apparatus Corp., 3831 Tyrone Blvd. N, St. Petersburg, FL 33709 (Model EC370) or Hoefer Scientific Instruments, 654 Minnesota St., Box 77387, San Francisco, CA 94107 (Model HE33). Best results are obtained with an 8-well, 1.5-mm-thick well former.

5 power supplies

2 water baths or dry baths, 37° C and 65° C, each with a cast aluminum micro test tube block (Thermolyne)

5 adjustable micropipetters, 10–100 µl (e.g., Eppendorf Digital Pipette)

Supplies

5 micro test tube racks	Thermometer
Pipet tips	Aluminum foil
Ziploc (Dow) sandwich bags	Disposable gloves
Boiling-water bath	5 permanent marking pens

5 French curve templates and rulers

Polypropylene micro test tubes with attached caps, 0.5 ml

5 staining dishes with covers, approximately 100 × 75 × 25 mm

Reagents

1 liter TAE running buffer (0.05 M Tris–0.002 M sodium acetate–0.03 N acetic acid–0.002 M EDTA, pH 7.8). To a 1000-ml volumetric flask, add 4.85 g tris(hydroxymethyl)aminomethane, 1.64 g sodium acetate (anhydrous), 1.7 ml glacial acetic acid, 0.74 g ethylenediamine tetraacetate, disodium

salt·2H$_2$O. Bring the volume to 1000 ml with distilled water and then adjust the pH to 7.8 with 10 N NaOH.

100–125 ml of 1% (w/v) agarose (e.g., BRL cat. no. 5510UA) in TAE running buffer. The volume required for a 3-mm-thick gel is generally 20–25 ml, depending on the particular apparatus used. Before casting the agarose gel, the ends of the casting deck must be sealed. Depending on the system, this is done with slabs of cork or Teflon (DuPont) tape. The well former is set near one end of the casting deck and the entire unit is placed on a level surface. (The unit must also be level during the electrophoresis.) The agarose is dissolved in the running buffer by heating it in a boiling-water bath until it is *completely* dissolved. When the temperature of the agarose solution has cooled to 50° C, it is poured into the casting deck. When the gel has completely set (~20 min), the well former and damming walls/Teflon tape are removed. The gel should be positioned within the electrophoresis chamber so that the sample wells are closer to the cathodic buffer reservoir. The reservoirs are then filled with running buffer until the surface of the gel is covered with 1–2 mm (no higher!) of buffer. The gel can be cast any time up to 6 hr prior to the lab session, provided that the gel slab is kept covered with running buffer.

200 ml working solution of Stains-all. Prepare a 0.1% stock solution by adding 75 ml formamide to 75 mg Stains-all (Kodak Laboratory Chemicals) in a 125-ml Erlenmyer flask wrapped in aluminum foil. Dissolve the stain by stirring with a magnetic stirring bar for 4–6 hr. Working solution: shortly before use, mix 10 ml stock solution, 90 ml formamide, and 100 ml distilled water, in that order. The working solution should be discarded after use, but the stock solution is good for several weeks, stored in the dark at 4° C.

500 µl "stop" solution, dispensed in 5 micro test tubes. Add 372 mg Na$_2$EDTA·2H$_2$O to 4.9 ml double-distilled water. Using narrow-range pH paper (do not immerse the paper in the liquid), adjust the pH to 7.5 by adding dropwise 5 N NaOH. When the Na$_2$EDTA·2H$_2$O is dissolved, add 100 mg sodium dodecyl sulfate and 10 mg bromphenol blue. When these are dissolved, add 5.0 ml glycerol with a 10-ml pipet and mix well by drawing the liquid up and down repeatedly in the pipet, using a rubber bulb pipet filler. Store at 4° C.

250 µl *Hind*III, 10 U/µl (BRL cat. no. 5207SA), dispensed with sterile pipet tips in 5 micro test tubes

250 µl *Xho*I, 10 U/µl (5231SA), dispensed with sterile pipet tips in 5 micro test tubes

250 µl 10× Reaction Buffer 2 (supplied with above enzymes), dispensed with sterile pipet tips in 5 micro test tubes

250 µl lambda DNA, dispensed with sterile pipet tips in 5 micro test tubes. BRL lambda DNA (cat. no. 5250SA) is supplied at concentrations of 250–600 µg/ml, depending on the particular lot. Dispensed within the concentration range of 400–600 µl/ml, the λ DNA should give excellent results. To use a batch with a lower concentration, have students increase the volume of the λ DNA and decrease the volume of distilled water in each micro test tube.

2.0 ml double-distilled water, dispensed in 5 micro test tubes

PROJECT **18**

This project requires two lab sessions to complete. During the first lab session, students prepare the digests, electrophorese the samples, and place the gel in the stain. The bands are visible in several hours, though overnight staining is recommended. Accordingly, the following morning, either a team member or the instructor/lab assistant should remove the gel from the stain and do the destaining.

In the second lab session, students can examine the gel and collect/analyze their data. The gels will last months if stored in the refrigerator in a sealed plastic bag wrapped with aluminum foil. So that the enzymatic treatments and electrophoresis can be completed in a single lab session, the agarose gels must be cast beforehand. Instructions for casting the gel and setting up the electrophoresis chamber are given in the Reagents section for Project 17, page 309.

With the exception of restriction enzymes and phage λ DNA, the materials required for this project are the same as those listed for Project 17. Also required for Project 18 are the items listed below. Students should work in teams of four; quantities are calculated for 20 students working in five teams. The *Eco*RI and 1 Kb DNA Ladder are available from Bethesda Research Laboratories (BRL), Gaithersburg, MD 20877. *The electrophoresis equipment should always be supervised because of the potential shock hazard associated with the power supply.*

Equipment

5 adjustable micropipetters, 2–10 μl (e.g., Eppendorf Digital Pipette)

Reagents

250 μl *Eco*RI, 10 U/μl (BRL cat. no. 5202SA), dispensed with sterile pipets in 5 micro test tubes

250 μl 10× Reaction Buffer 3 (supplied with the *Eco*RI), dispensed with sterile pipets in 5 micro test tubes

125 μl of the 1 Kb DNA Ladder (BRL cat. no. 5615SA), dispensed with sterile pipets in 5 micro test tubes

250 μl calf thymus DNA, 100 μg/ml, dispensed in 5 micro test tubes. Add 25 mg calf thymus DNA (Sigma Chemical Co., cat. no. D1501) to 25 ml standard saline citrate (SSC), pH 7.0 (see p. 296) at 4° C. Keep refrigerated and mix periodically over several days, until the DNA goes into solution. Aliquots can be stored almost indefinitely at −20° C.

PROJECT **19**

For the C-banding, students should be provided with unstained slide preparations from short-term human leukocyte cultures harvested about 1 week prior to the lab session in which the C-banding will be done. A total of three cultures should suffice for a class of 20 students since each harvested culture yields 9 to 10 slides. For all cultures, the blood should come from a healthy individual (other than a student) known to have a normal karyotype. The donor should remain anonymous.

The instructions for initiating and incubating the blood cultures and for harvesting the cells follow. Care should be taken when handling blood since there is always the slight possibility that the blood is contaminated with infectious agents. Accordingly, disposable gloves should be worn, spills wiped up, the area disinfected, and waste materials disposed of properly. It is advisable that the individual doing the harvest be the one who initiates the culture with his/her own blood.

1. Reconstitute the lyophilized chromosome medium with the supplied reconstituting fluid and place in a 37° C incubator for 15 min.
2. Use aseptic technique to obtain the blood sample. After washing the hands with soap and water, and drying them, clean the tip of the ring or middle finger with a sterile alcohol swab. When the fingertip has dried, lance it with a sterile lancet, "milk" the finger, and collect three heparinized capillary tubes of blood (four to five drops). Each capillary tube is filled by holding it nearly level and touching the tip to the drop. Remember to clean the lanced fingertip with alcohol and to discard the lancet, which should never be used more than once.

3. Using a capillary dropper bulb, transfer the blood to the reconstituted medium. Invert the capped culture several times to mix the contents. Record the date and time that the culture was started.

4. Incubate the culture at 37° C for 66–72 hr. Twice each day, invert the culture several times to dislodge the cells from the bottom and disperse the culture.

5. Four hours before the incubation is finished, add 1.0 ml of the colchicine solution* to the culture. Invert the capped culture several times before returning it to the incubator.

6. After the 4-hr incubation with colchicine, disperse the culture by inverting the vial several times. Then, with a Pasteur pipet, transfer the entire culture to a 15-ml conical centrifuge tube. Centrifuge at 500 g for 5 min.

7. With the Pasteur pipet, carefully aspirate and discard almost all of the liquid above the pellet. Add 5 ml of warm (37° C) Hanks balanced salt solution, resuspend the cells with the Pasteur pipet, and centrifuge at 500 g for 5 min. Whenever resuspending, continue until all clumps are dispersed.

8. With the Pasteur pipet, carefully aspirate and discard almost all of the liquid above the pellet. Add 5 ml of 0.075 M KCl, 24° C, and gently resuspend the cells with the Pasteur pipet. Allow the cells to remain in the hypotonic solution for 12 min.

9. Centrifuge at 200 g for 5 min. While the cells are being centrifuged, prepare the fixative as follows. Add 45 ml methanol and 15 ml glacial acetic acid to an Erlenmyer flask. Mix well and keep the flask capped.

10. With the Pasteur pipet, carefully aspirate and discard almost all of the liquid above the pellet. Then, carefully draw up all of the pellet (cells) into the lower, tapered portion of the Pasteur pipet. Place the pipet aside.

11. Add 5 ml of fixative[†] to the nearly empty centrifuge tube. Place the tip of the Pasteur pipet (containing the cells) just beneath the surface of the fixative and slowly compress the pipet bulb so that the cells run down the side of the centrifuge tube and settle at the bottom. The dark-brown color is due to the action of the fixative on the hemoglobin released from the hemolyzed erythrocytes. The pellet will progressively whiten with subsequent changes of the fixative.

12. Gently resuspend the cells with the Pasteur pipet. Cover the tube with Parafilm and allow the cells to fix for 15 min. The swollen, hypotonic-treated cells are rather fragile and should be handled very gently.

13. Centrifuge at 200 g for 5 min.

14. With the Pasteur pipet, carefully aspirate and discard almost all of the fixative above the pellet.

15. Add 4 ml of fresh fixative and gently resuspend the cells with the Pasteur pipet.

16. Centrifuge, aspirate, and resuspend the pellet in 4 ml of fresh fixative, as described in steps 13, 14, and 15.

17. Again, centrifuge, aspirate, and resuspend as described in steps 13, 14, and 15, but this time gently resuspend the pellet in only 0.3–0.4 ml of fresh fixative.

18. Make the air-dried preparations on alcohol-cleaned slides that have been stored in chilled, distilled water in Coplin jars. Wipe the water from the bottom of the slide and then with the Pasteur pipet held 6–10 cm above the slide, let two or three drops of the cell suspension drop onto the horizontal, wet slide surface. Blow vigorously on the surface and immediately place the slide on a hot plate until almost dry (about 10 sec). The temperature of the hot place should be set to maintain the temperature of a 50-ml beaker of water at 50°-55° C.

19. Store the unstained slide preparations at low humidity in a lightproof slide box.

*CAUTION: *Colchicine is poisonous; do not pipet by mouth.*
†CAUTION: *The fixative is poisonous; do not pipet by mouth.*

On the day that the C-banding is to be done, immerse the slides in 0.2 N HCl, room temperature, for 30 min. Rinse the slides in several changes of distilled water and allow to air dry. Students thus begin with slides that are ready for the Ba(OH)$_2$ treatment. For the C-banding, each student should work individually, but reagents should be shared by each pair of students. The items required for the C-banding are listed separately from the items required for the cell culture and harvest, and quantities are calculated for 20 students working in 10 teams.

Equipment for Cell Culture and Harvest

1 37° C incubator	1 clinical centrifuge

1 hot plate set to maintain temperature of a 50-ml beaker of water at 50°-55° C

Supplies for Cell Culture and Harvest

1 50-ml graduated cylinder	Disposable gloves
1 1-ml pipet	Sterile alcohol swabs
2 5-ml pipets	1 pair forceps
1 pipet filler, rubber bulb type	Parafilm
1 test tube rack	Kimwipes

3 15-ml graduated, concial centrifuge tubes

2 125-ml Erlenmyer flasks with screw caps

Microscope slides with frosted end. New slides from the box should be cleaned with 95% ethanol and stored in cold distilled water in Coplin jars.

20 Giemsa-stained slide preparations of human chromosomes, available from Carolina Biological Supply Co. or Ward's Natural Science Establishment, Inc.

Reagents for Cell Culture and Harvest

2 Difco TC Chromosome Microtest Kits, cat. no. 5060-32-8. Each kit contains material for two cultures: two vials each of lyophilized culture medium, reconstituting fluid, arresting solution (colchicine), and Hanks balanced salt solution. Each kit also contains sterile blood lancets, Pasteur pipets with bulbs, sterile capillary tubes (for collecting blood) with bulbs, and Giemsa stain. (Difco products are available from Fisher Scientific and other supply companies.)

50 ml 0.075 M KCl: dissolve 5.59 g KCl in distilled water to make 1000 ml of solution

45 ml reagent-grade methanol from an unopened bottle

15 ml reagent-grade glacial acetic acid from an unopened bottle

Equipment for C-Banding

2 50°C water baths	1 60°C oven

20 compound microscopes equipped with cross-hair reticles

Supplies for C-Banding

20 Coplin jars	Filter paper
2 50-ml graduated cylinders	20 pairs forceps
2 pipet fillers, rubber bulb type	2 2-ml pipets
Disposable gloves	Kimwipes

Pasteur pipets, 146 mm, and dropper bulbs

10 square Petri dishes, 100 mm^2 × 15 mm deep

40 rubber washers, diameter approximately 25 mm (for supporting the slides in the moist chambers). Alternatively, glass rods can be used.

Coverslips, 24 × 60 mm (for the 2×SSC incubation) and 22 × 50 mm (for mounting with Permount)

Reagents for C-Banding

200 ml 0.2 N HCl dispensed in 5 Coplin jars

200 ml 5% (saturated) aqueous solution of barium hydroxide dispensed in 5 Coplin jars. To each Coplin jar, add 2.0 g Ba(OH)$_2$·8H$_2$O and 40 ml distilled water. Mix well. Just before the slides are immersed in the solution, the surface will be cleared of scum so that slides will not pick up Ba(OH)$_2$ particles as they are withdrawn from the solution.

500 ml 2×SSC solution (0.3 M NaCl–0.03 M sodium citrate): dissolve 8.77 g NaCl, 4.41 g Na$_3$citrate·2H$_2$O in distilled water to make 500 ml of solution. Adjust pH to 7.0 with 1 N HCl. Prepare within 1 week of use.

500 ml 0.01 M phosphate buffer, pH 6.8: dissolve 0.35 g Na$_2$HPO$_4$, 0.34 g KH$_2$PO$_4$ in distilled water to make 500 ml of solution. Adjust pH to 6.8 with 1 N NaOH.

25 ml Harleco Giemsa Blood Stain, Original Azure Blend Type, available from Baxter Scientific Products, 100 Raritan Center Parkway, Edison, NJ 08818.

175 ml xylene dispensed in five Coplin jars

50 ml Permount dispensed in five glass-rod dropping bottles

PROJECT 20

For the Hoechst–Giemsa staining, students should be provided with unstained slide preparations from short-term human leukocyte cultures harvested up to one week prior to the lab session in which the Hoechst-Giemsa staining will be done. A total of three cultures should suffice for a class of 20 students since each harvested culture yields 9 to 10 slides. For all cultures, the blood should come from a healthy individual other than a student, and no attempt should be made to karyotype the cells. The donor should remain anonymous.

The instructions for initiating and incubating the blood cultures and for harvesting the cells are given on pages 310–311; the two additional steps listed here are also required for this project. Care should be taken when handling blood since there is always the slight possibility that the blood is contaminated with infectious agents. Accordingly, disposable gloves should be worn, spills wiped up, the area disinfected, and waste materials disposed of properly. It is advisable that the individual doing the harvest be the one who initiates the culture with his/her own blood.

1. Wrap a piece of aluminum foil around the culture vials so that they will not be exposed to light during the incubation period.

2. Add 0.1 ml of the BrdU solution* to each reconstituted culture prior to adding the blood.

The equipment, supplies, and reagents required for initiating and harvesting three leukocyte cultures are listed on page 312. Required items for the BrdU–Hoechst–Giemsa procedure are listed below.

Equipment and Supplies

20 compound microscopes   Coverslips, 22 × 50 mm

2 2-ml pipets   10 Coplin jars

*CAUTION: *BrdU is toxic; do not pipet by mouth.*

2 10-ml pipets

2 pipet fillers, rubber bulb type

2 50-ml graduated cylinders

3 fluorescent lamps, each with two cool-white bulbs

Pasteur pipets, 146 mm, and dropper bulbs

Draining slide holders

20 pairs forceps

Aluminum foil

Benchkote

Reagents.

31 ml 2×10^{-3} M BrdU in Hanks balanced salt solution: dissolve 19 mg 5-bromo-2'-deoxyuridine* (Sigma Chemical Co.) in 31 ml sterile Hanks solution (Difco Laboratories). The solution can be filter-sterilized or used as is, since there are antibiotics in the culture medium. The addition of 0.1 ml to a 4-ml culture will give a final BrdU concentration of 5×10^{-5} M.

160 ml aqueous solution of Hoechst 33258, 200 μg/ml, dispensed in four Coplin jars: dissolve 32 mg Hoechst 33258[†] in 160 ml distilled water. The dye is available from Sigma Chemical Co., Catalog No. B2883.

250 ml McIlvaine's buffer, pH 7.0: dissolve 0.85 g citric acid, anhydrous (0.93 g citric acid, monohydrate), 5.85 g Na_2HPO_4 in distilled water to make 250 ml of solution.

500 ml 0.01 M phosphate buffer, pH 6.8: dissolve 0.35 g Na_2HPO_4, 0.34 g KH_2PO_4 in distilled water to make 500 ml of solution. Adjust pH to 6.8 with 1 N NaOH.

25 ml Harleco Giemsa Blood stain, Original Azure Blend Type, available from Baxter Scientific Products, 100 Raritan Center Parkway, Edison, NJ 08818.

175 ml xylene dispensed in five Coplin jars

50 ml Permount dispensed in five glass-rod dropping bottles

PROJECT **21**

The following preparations are recommended for photomicrography: Wright- or Giemsa-stained blood cells, mitotic stages in root-tip squashes of *Vicia faba*, and meiotic stages in testis squashes of the grasshopper.

Listed are the basic items required for photomicrography, film processing, and printing. A darkroom is required only for printing; Technical Pan Film 2415 can be loaded in a developing tank using a changing bag. For the latest, most complete information on any Kodak or Polaroid product, always refer to the manufacturer's instructions. For technical assistance, there are also toll-free numbers for Kodak (800-242-2424) and Polaroid (800-225-1618).

Equipment

Trinocular microscopes equipped with 35mm and Polaroid attachment cameras, and with exposure meters

Enlarger

Polaroid 35mm AutoProcessor

Supplies

Stage micrometer slide

Kodak Projection Print Scale

Scissors

Photographic timer

Print tongs

Squeegee

*CAUTION: *Wear gloves and mask when weighing out BrdU.*

[†]CAUTION: *Wear gloves and mask when weighing out Hoechst 33258.*

Film clips   Paper cutter
Glassine envelopes   Tacking iron
Polaroid 35mm Slide Mounter   500-ml graduated cylinder
Polaroid 35mm Slide Mounts   600-ml beaker
Enlarger easel   Funnels
Enlarger focusing scope   Thermometers
Camel's hair brush   Changing bag (optional)
Kodak Dry Mounting Tissue, Type 2
Kodak Polycontrast II filters, grades 1–4

Kodak Technical Pan Film 2415. Available in magazines with 36 exposures, as well as in 150-foot rolls for economical bulk loading. Unexposed film can be stored almost indefinitely in the refrigerator or freezer. Allow 2–3 hr for refrigerated film to reach room temperature before opening the package; for frozen film, allow 5 hr.

Bulk film loader and reusable 35mm magazines (optional)

Kodak Polycontrast III RC Paper, F (glossy), medium weight. Unexposed print paper can be stored almost indefinitely in the refrigerator or freezer. Allow it to come up to room temperature before opening the package.

Polaroid Type 667 (8 exposures) and PolaPan CT (12 and 36 exposures). Store in refrigerator and allow 2–3 hr for the film to reach room temperature before opening the package. *Do not freeze.* Each roll of PolaPan comes complete with a processing pack.

Filters: Assorted neutral density filters (50%, 25%, 12.5% transmittance); Wratten filter No. 58 and/or No. 11. Wratten gelatin filters are easier to handle and more durable if cut and then mounted in 2 × 2-in. glass projection slides.

Plastic 35mm developing tank (Paterson or Star-D, for example)

Safelight with 15-watt bulb and OC filter

Trays for developer, stop bath, and fixer

Reagents

All the photgraphic chemicals listed are toxic if swallowed or inhaled, and repeated contact may cause skin irritation and allergic skin reaction.

Kodak HC-110 developer. Make up the stock solution according to package directions and then dilute just before use. Dilution F (for moderate contrast) is given in the project. If high-contrast Technical Pan negatives are desired, use dilution D (1:10): mix 30 ml HC-110 stock solution + 270 ml water; development time is 6 min at 20° C. With dilution D, Technical Pan has an ISO rating of 100.

Kodak Indicator Stop Bath. It can be used for many rolls of film and many prints until the solution turns blue.

Kodak Dektol developer. One gallon has a capacity for 100 8 × 10-inch prints.

Kodak Fixer. One gallon has a capacity for 100 8 × 10-inch prints.

Kodak Photo Flo, diluted as per package directions

Kodak Hypo Clearing Agent (optional)